TryTank: Innovation That Serves the Gospel

**The Church doesn't need another program.
It needs a place to experiment.**

At TryTank, we turn bold ideas into working models for ministry.
We've helped the church grow lay-led communities, evaluated national initiatives, and tested over 90 ways the Church can meet the moment—quickly, faithfully, and with data-driven insight.

Now we're helping ministries like yours do the same.

Our Services

- **Evaluation Impact Reports (EIRs):** measure what's working and tell your story to funders.
- **Innovation Cohorts:** hands-on training to design, test, and scale new ministry ideas.
- **AI Tools for Ministry:** ethical chatbots, analytics, and digital formation resources built for faith-based organizations.
- **Workshops & Retreats:** customized experiences to help leaders think creatively and act courageously.

Every project begins with a question worth asking and ends with insight you can use.

Because faithful leadership isn't about predicting the future—
it's about prototyping it.

Learn more or start a conversation at **TryTank.org**

TryTank
Research Institute

Praise for Church Leader's Guide to Artificial Intelligence

Dr. Christopher Benek has written a prophetic and practical guide for one of the Church's most defining moments in history. He is helping us see that AI is not simply a technological issue—it is a theological one. With remarkable clarity and pastoral wisdom, Benek invites leaders to move beyond fear and fascination, toward fruitful engagement.

What I appreciate most is that this book does not position the Church as a spectator to technological change, but as a shaping voice within it. Christopher combines a deep grasp of emerging technologies with a rich theological imagination, reminding us that human dignity, not digital efficiency, must remain at the center of all innovation.

This is more than a manual on AI—it's a call for the Church to think redemptively about culture, vocation, and the future of humanity. If you care about the Gospel's witness in the modern world, read this book carefully.

— Alan Platt, Founder, Doxa Deo & City Changers Movement

"Chris Benek's book arrives at a pivotal moment — just as we surrender more of our choices to AI. It reminds all of us, but especially the leaders among us, that true progress shouldn't just be measured in profit or convenience, but in wisdom, dignity and community. In the AI-driven Cognitive Era, we must not only build smarter systems, but lead with empathy and love, design with purpose and respect, and protect the empathy and wonder that make us human."

— Olaf Groth, PhD; Faculty at UC Berkeley Haas; CEO at Cambrian.ai; author of Solomon's Code and The Great Remobilization

---

More and more, the church is finding itself perfectly prepared to do ministry in a world that no longer exists. This book is written for ministry leaders, like me, who may know next to nothing about artificial intelligence but who, like me, have a sneaking suspicion that it will increasing matter for future of the church. The Church Leaders Guide to Artificial Intelligence sets framework for understanding the rapidly evolving world of AI, all through a theological lens to help us name both the gifts and the risks of AI.

— Mark DeVries, Founder of Ministry Architects, Co-Founder of Ministry Incubators

---

The Rev. Dr. Christopher Benek offers the most accessible, theologically grounded introduction to AI and the Church I have yet seen. Clear without being simplistic, this guide invites pastors, educators, and theologians alike into practical discernment rather than fear or fascination. Rooted in Scripture and pastoral experience, it equips the Church to engage emerging technologies as sites of vocation and witness. Benek's voice is generous, wise, and deeply grounded in faith's intellectual and ethical traditions.

— The Rev. Dr. Michael W. DeLashmutt, Senior Vice President, Associate Professor of Theology, and Dean of the Chapel, The General Theological Seminary

---

Church Leader's Guide to Artificial Intelligence:

Understanding AI's Impact on Ministry, Theology, and Culture

Rev. Dr. Christopher Benek

TryTank
PRESS

Alexandria, VA

Copyright © 2025 by Christopher Benek. Published 2025.
20 19 18 17 16 15 1 2 3 4 5 6 7 8 9

TryTank
PRESS

ISBN: 978-1-971082-00-4.

Cover and design: Richard Oliver

Printed in the United States of America

Stay Ahead. Lead the Change.

Church Leader's Guide isn't just a book—it's a movement of pastors, innovators, and faithful pragmatists re-imagining ministry in real time.

Each edition opens new ground, but the conversation keeps evolving.

The Church Leader's Guide E-Newsletter brings you:

- Fresh insights on technology, theology, and culture
- Practical tools you can use this month in your ministry
- Short reflections from trusted voices shaping the future Church
- Early access to new guides, studies, and TryTank experiments

Join the growing community of leaders who believe the Church's best days aren't behind us—they're being built right now.

> ### Subscribe to our newsletter today at
> ### ChurchLeadersGuide.org

Because faithful leadership never stands still.

Church Leader's Guide
A TryTank Press Series
Learn. Adapt. Lead Forward.

Foreword

Throughout my forty-plus years in ministry, I've navigated profound shifts in culture, communication, and economics. I've also learned that disruption, though uncomfortable, is often the catalyst God uses to prepare His people for what's next. In every generation, faithful leaders have faced the same question: Will we resist change and preserve the familiar, or will we adapt, innovate, and lead forward in faith?

With *The Church Leader's Guide to Artificial Intelligence*, my good friend Christopher Benek helps us confront the next great disruption head-on. What he offers is neither hype or fear, but a clear-eyed, pastoral framework for understanding how this technology will impact ministry, theology, and culture—and why the Church must lead, not lag, in this moment.

I first met Christopher about ten years ago, during my own search to understand, discover, and forecast how artificial intelligence might affect the future of the local church. Even then, when few ministry leaders were paying attention, he was already doing the theological and practical work—speaking to engineers, ethicists, and pastors with equal fluency. Then and now, I believe he stands among the world's foremost clergy experts on AI. As well-versed as he is in the ethics, implications, and trajectory of what he calls "Alternative Intelligence," what I appreciate even more is his pastoral heart and desire to see committed followers of Jesus at the forefront of AI engagement, usefulness, and humanity.

The Church has been here before. The printing press democratized access to Scripture. The industrial revolution redefined vocation. The internet reshaped communication and community. Each wave of change threatened existing structures, yet each also carried within it divine opportunity. The same is true of artificial intelligence. The question is not whether AI will change the Church—it already is—but whether the Church will discern and direct that change toward redemptive ends.

I coined the term *Church Economics* to describe how pastors in the twenty-first century must learn to leverage church assets to expand missional vibrancy and pursue financial sustainability. This book extends that same spirit of innovation. Where *Church Economics* reframes the stewardship of resources, Christopher's work challenges us to rethink the stewardship of intelligence itself—human, artificial, and spiritual. In both cases, the goal is the same: to align church assets with the needs of a changing world without compromising the gospel's message and mission.

If you've read any of my writing, you know I believe the Church must move beyond incremental improvements to embrace disruptive innovation. That phrase, borrowed from business theory, captures what Jesus often did in practice. He upended religious systems, overturned tables, and called ordinary people into extraordinary purpose. True discipleship has always been disruptive because it calls us to see differently, live differently, and lead differently.

Christopher approaches AI through that same lens. He neither idolizes technology nor demonizes it. Instead, he invites leaders to bring a theological imagination to the table, to see AI not merely as a tool to be used but as a terrain to be stewarded. His writing blends the rigor of a scholar with the heart of a pastor, reminding us that what's at stake is not just our ministries but our humanity.

Reading this guide, you'll recognize situations familiar to your own context: pastors overwhelmed by administrative tasks, churches struggling to reach digital generations, congregations yearning to remain relevant even as they age. AI touches each of these realities. It can help us translate Scripture faster, analyze community needs more precisely, and communicate the gospel more broadly. Yet as Christopher reminds us, it cannot replace the incarnational presence that defines Christian ministry. AI may amplify our capacity, but only the Spirit transforms the heart.

For decades, I've argued that the American Church must awaken to new cultural and economic realities, learning to multiply revenue, repurpose facilities, and reimagine its role as an engine of spiritual and social renewal. What *The Church Leader's Guide to Artificial Intelligence* adds to that conversation is another dimension: the stewardship of knowledge, creativity, and discernment in a world increasingly mediated by machines. In that sense, this book belongs to the same stream of thought that has driven the multiethnic church movement, the emerging field of *Church Economics*, and the theology of innovation.

It is not a technical manual; it is a prophetic guidepost for vocational and lay leaders alike. The task before us is not simply to adopt AI tools but to disciple people who can use them faithfully. It is to teach a generation how to discern the difference between data and truth, between algorithmic prediction and divine revelation.

Throughout this insightful read, Christopher demonstrates that the image of God in humanity is not diminished by technology but reaffirmed through wise stewardship of it. His call reflects the spirit of every reformer and revivalist who refused to let fear dictate the future. He invites us to be both childlike and courageous: open to learning yet anchored in eternal truth.

When I planted the Mosaic Church of Central Arkansas in 2001 and later launched the Mosaix Global Network in 2004, I could never have imagined the pace of change we'd witness by 2025. Yet here we stand: churches leveraging facilities to generate income, nonprofits operating as mission arms, algorithms shaping our communication, and AI co-writing worship sets. In such a world, the local church must again become a laboratory for innovation, a model of ethical imagination, and a witness to what technology can look like when surrendered to the Lordship of Christ.

This book is a gift to that future. Read it slowly, discuss it deeply, and let it stretch your imagination for what faithful leadership looks like in the age of intelligence. History shows that those who embrace disruption with discernment don't just survive the future. They shape it.

Dr. Mark DeYmaz, D.Min.
Directional Leader, *Mosaic Church of Central Arkansas*
Co-founder & Directional Leader, *Mosaix Global Network*
Author of *Disruption: Repurposing the Church to Redeem the Community and The Coming Revolution in Church Economics*

Contents

CONTENTS

Introduction

Instead of trying to produce a program to simulate the adult mind, why not rather try to produce one which simulates the child's?

~ Alan Turing, mathematician and pioneer of computer science

Why AI Matters for the Church

On Monday morning, before your first cup of coffee, AI has already shaped your congregation. A teen in your youth group woke up to an algorithm-curated feed. A deacon drafted a challenging email with the help of an AI assistant. A congregant applied for a job, and an unseen hiring algorithm filtered their résumé before any human ever saw it. Another member asked an AI private questions they're afraid to ask a pastor.

By Sunday morning, the people in your pews distinctly carry a week's worth of AI's imprint. The high schooler, shaped by hours of TikTok and Instagram reels, arrives anxious and thin on hope. The deacon's hard-to-write email - drafted with AI - read competent but not compassionate. And now they seek advice on how to pivot their approach. The jobseeker whose résumé an algorithm screened has shown up asking for prayer and help getting a human to actually read it. The member who asked private questions of a chatbot comes seeking embodied, Spirit-led care from a church leader.

Whether you serve a house church or a multi-site ministry, an online community or a multilingual neighborhood congregation, shepherding now happens in a world quietly mediated by machine intelligence. That isn't a future problem - it's ministry today. Ignore AI's impact, and you'll miss the practical needs your congregants bring; fail to look ahead, and you'll be planning for a world that's already passed.

The purpose of this book is to help leaders in the church understand the impact of AI on ministry, theology, and culture by providing an accessible, ministry-focused guide. The goal is to help leaders navigate AI's emerging challenges and opportunities within a Christian context. And the keyword in that last sentence is *emerging*.

Understanding AI's impact on ministry, theology, and culture is not a fixed target. It is an exponentially evolving one. And because of that, I fully confess that this book isn't complete. In fact, it's not meant to be. Instead, it's meant to be a living document that will be updated over time. This is why it will intentionally remain largely print-on-demand. That way, with your help, it can be updated as AI changes and as we change with AI – and as we continue to evolve beyond printed books.

Why You Matter on This Issue

I say "with your help" because that is what this guide is – an invitation to help – in arguably the most significant technological shift in human history.

In 2018, I was invited to speak in Japan at the International Conference of Social Robotics near Tokyo. I was the first clergy member in history to receive this invitation—and the only one present.

When I arrived, a large group session was underway, with clusters of roboticists from around the world gathered at tables. I observed as the entire group watched a video of a robot interacting with humans in a public setting. Following the video, everyone broke out into their respective small groups to discuss what they had just witnessed.

My host invited me to sit at one of the tables and join in the conversation. The assignment was for each group to discuss what they felt was awkward about the human/robot interaction that they had just witnessed. One by one, each person at the table offered their opinion. Then, it was my turn. When I finished, a young woman asked, "Who are you? I know who everyone in this room is, except for you. And if I don't know who you are, how do you know so much about this topic?"

When I replied, "I am a pastor from Florida," they were all genuinely shocked. However, I explained that what I do requires me to regularly evaluate what is really happening when a person's behavior and actions don't match their words. And so, as a result of a lot of formative practice, looking at where human-robot interaction was "off" came pretty easily to me.

I tell you this story to emphasize a point about you. Some of the best voices I've heard on issues of emerging tech are from people who don't seem, at face value, to have much to do with emerging tech. But they have other skill sets that dramatically impact the discussion about such technologies. As a church leader, you may not be an AI expert, but, more than likely, you are an expert at dealing with people. You know your tribe, their context, the not-yet-measured nuances that make them them, maybe better than anyone. You know what to do when the data suggests one thing, but the situation is unfolding in another way—a kind word, a firm boundary, or a targeted prayer that infuses hope. This is why I believe God has formed you for this precise moment in human history, and God can use you in it.

The Church's gifts—attention, discernment, empathy, moral imagination— belong in the rooms where AI is being designed, deployed, and debated. You may not be an engineer, but you are an expert in caring, communities, and the slow wisdom of formation.

This guide is an invitation to bring your knowledge and discernment to the AI conversation. That is because, as a church leader, you do not need to become a computer scientist to lead well. You just need a clear grasp of the stakes, a basic shared vocabulary for your teams and elders, and a theological compass that holds steady when the headlines lurch between awe and alarm. You may be a novice in the AI arena, but you are an expert at people.

The aim here is not to journey down the rabbit hole, sprint to the far edges of speculation, nor to drown you in jargon. The goal is to get you to put your toe in the water, to name the currents that already tug at your people, and then to equip you to wade in wisely. There are entire libraries devoted to the secular applications of AI, and for every domain we just briefly note in this volume, many more will not be addressed. That is intentional. *The goal is not to say everything,* but to say enough that your own wheels start turning for your context.

How This Book Will Help You

The aims of this book are straightforward. **Part 1** provides a grounded understanding of what AI is and how it operates in everyday life. You will meet the core ideas without myth or mystique, and you will see how the Church's wealth of experience can ground and strengthen our approach today. **Part 2** turns to ministry practice—worship, discipleship, missions, stewardship, and care—showing where AI can serve as a tool and where its limits must be named for the sake of human dignity and spiritual depth. **Part 3** engages the ethical and theological questions that cannot be avoided: identity, autonomy, bias, governance, work, warfare, creation care, and the possibility of systems that one day exceed our own capacities. In Part 4, we consider the potential for the Church's proactive role in shaping AI's development and application, from public witness and policy engagement to mentoring the next generation of Christian technologists and ethicists. Along the way, you will find language you can use with your board, your staff, your small groups, and even your teenagers, along with practices you can begin to act on this month.

None of this replaces the incarnational heart of ministry. Technology can perform tasks more swiftly and efficiently for you; it cannot love your people for you. It can summarize a biblical commentary; it cannot sit beside a hospital bed. It can help you see patterns in your congregation's needs in the collected data; it cannot pray in your voice or weep with those who weep. Our calling is to let tools be tools and to keep persons, not efficiencies, at the center. That means pursuing transparency when we use AI, guarding privacy with rigor, correcting bias with justice, and measuring success not by speed alone but by faithfulness to Christ and the flourishing of our neighbors.

Turing's childlike mind is a fitting image for our posture in dealing with AI. The Church approaches this era as a learner—curious, teachable, resilient, unafraid to ask basic questions and unwilling to pretend certainty where there is none. Childlike does not mean naïve; it means open to growth, quick to repent, eager to try again. It means we can hold whimsy and gravity together, because hope and humor often keep courage alive when the challenges are immense. The road ahead will be unpredictable. A gentle, imaginative spirit will help us face the journey without fear.

If you are holding this book in your hands because someone in your church asked a question you could not answer, or because your inbox is filling with digital tools you do not trust, or because you sense a cultural shift you need to shepherd through, you are in the right place. Take a breath. Keep your Bible open. Pray as you plan. Teach what you learn. And above all, remember the center: the One who is the same yesterday, today, and forever. AI will change many things. It does not change the call to love God and neighbor. It does not change the resurrection hope that anchors us. It does not change the Church's mission to speak truth, do justice, and walk humbly. With that center firm, we can learn quickly, lead wisely, and— together—help this technology serve the Gospel rather than supplant it.

This book, therefore, is an invitation to co-create its future. As the field evolves, editions will be updated, examples revised, cautions revised, and practices refined. Your feedback, stories, and questions will shape those updates. In that sense, this is not just my guide; it is ours. As with every good pastoral work, it lives in conversation with the people it serves. It is a living document, and this is just its start.

Part 1:
Understanding AI Basics

Chapter 1:
What is Artificial Intelligence?

~ John McCarthy
computer scientist and father of
Artificial Intelligence[1]

What is Artificial Intelligence?   That is the question that everyone wants answered. The technology is still developing. The name is still inadequate. But what is certain is that Artificial Intelligence (AI) is one of the most transformative technological advancements of our age.[2]

Defining AI in Plain Terms

Not all automation is intelligent. At its core, the term Artificial Intelligence (AI) refers to a branch of computer science that focuses on creating systems capable of performing actions that require human-like intelligence and simulate human cognitive processes. Such actions include learning from experience, reasoning through complex problems, recognizing patterns, and understanding or generating human language. No longer a distant idea confined to science fiction, AI is now woven into the fabric of daily life. It shapes what we see on social media through recommendation algorithms, guides our conversations with voice assistants at home, curates the products and services we encounter online, and even supports large-scale industrial operations.

These systems can absorb vast quantities of data, identify meaningful patterns, and generate insights or solutions with a speed and accuracy far exceeding human abilities. In practice, this can mean identifying suspicious activity in financial transactions, recommending optimal medical treatments, or translating messaging instantly into multiple languages. AI's ability to interpret human language allows for the creation of conversational agents, translation tools, and educational platforms. It can also process images, audio, and video, enabling facial recognition, real-time transcription, and predictive text generation. In some specialized fields, such as chess, logistics, and specific areas of medical diagnostics, AI has already surpassed human performance to a significant extent.

While the term 'AI' is often used broadly, it is important for us to distinguish it from related concepts. Many tools still operate through rigid, preprogrammed rules and without adapting or learning. Machine learning is a subset of AI in which systems improve their performance by analyzing data rather than simply following fixed instructions. Deep learning is a further subset that uses neural networks—layered systems modeled loosely on the structure of the human brain—to process information.[3] These networks excel at recognizing patterns in images, speech, and text by passing data through many interconnected layers of "neurons," each refining the result.

Narrow, General, and Superintelligent AI

Today, AI manifests in several distinct forms. The most common is known as narrow or weak AI, designed to excel in a specific domain such as voice assistance, facial recognition, or spam filtering. This form of AI performs assigned tasks with remarkable proficiency, but cannot adapt far beyond its programmed purpose. The idea of general AI—Artificial General Intelligence (AGI)—machines capable of learning and reasoning across a wide range of activities at a human level—remains theoretical.

Still, some experts predict that such capabilities could be achieved by the middle of this century, with others suggesting that it may come even sooner. Beyond that lies the even more speculative concept of superintelligent AI, which would surpass human intellectual capacity in every dimension, including creativity, emotional insight, and strategic planning. This hypothetical future raises not only technological questions but also theological and existential ones about humanity's role in creation.

At its core, current AI is only as strong as the data it consumes. These systems do not generate wisdom out of nothing; they mirror the biases, assumptions, and limitations embedded in their training data. This reality makes conversations about truth, justice, and discernment all the more urgent.

Ancient Dreams, Modern Breakthroughs

Although the technical field of AI is relatively young, the desire to create intelligent, human-made beings is ancient. Greek mythology recounts the tale of Talos, the bronze guardian of Crete, and Hephaestus, the divine craftsman who forged golden servants with intelligence. Ancient Chinese and Indian texts include stories of mechanical beings and artificial life forms, while medieval legends describe scholars who fashioned talking heads or lifelike automata.[4]

During the Enlightenment, thinkers such as René Descartes even speculated whether humans themselves could be viewed as elaborate, divinely crafted machines.[5] In the book *Spiritual Ethics: Implications of Human Enhancement and Artificial Intelligence*, I argue that humans are themselves a kind of alternative intelligence in relation to God, and that, when considered through the lens of information theory, all matter can be understood as divine technology.[6]

The modern scientific pursuit of AI began in the mid-twentieth century, when figures like Alan Turing proposed that machines might one day "think" in a way that could be tested and measured.[7] From the initial optimism of the 1950s and 1960s, through subsequent periods of slow progress, the field advanced steadily until the breakthroughs of the 2010s in deep learning and neural networks ignited a new era of rapid innovation.[8]

Today, the capabilities of what we call AI are impressive. AI can analyze immense datasets in moments, automate tasks that once consumed hours of human labor, recognize intricate patterns invisible to the human eye, and generate works of music, literature, and art that mimic human creativity. AI can also adapt learning experiences to individual needs and expand accessibility.

Yet, even with these strengths, AI remains limited. Often struggling to understand the information it processes, AI, as far as we know, lacks self-awareness, empathy, and the moral capacity to make independent ethical judgments. It does not yet possess a moral code, advanced abstract reasoning abilities, or self-autonomy. Its reasoning is confined to the parameters set by its human creators, and its apparent intelligence is currently, in reality, a highly sophisticated form of pattern recognition rather than genuine comprehension. Though AI can appear to reason or create, it does not actually think as humans do. Its insights are simulations of intelligence, not the product of consciousness or moral imagination.

Promise, Limits, and Responsibility

Yet, still, the opportunities AI presents are significant. In healthcare, it can help detect diseases at earlier stages, tailor treatments to individual patients, and accelerate the development of new medicines. In education, it is able to make personalized learning possible on a scale once unimaginable.[9] And, in ministry, AI can aid in translating Scripture into multiple languages, streamline data regarding membership engagement, support outreach efforts by connecting digital seekers, and advance theological research by quickly retrieving and summarizing scholarly works.[10]

At the same time, these advancements bring serious challenges. Automation displaces jobs, requiring individuals and communities to adapt and reskill. AI-driven surveillance and mass data collection raise profound concerns about privacy and individual rights. Human biases embedded in training data can lead to AI-directed conclusions that result in unjust outcomes in hiring, law enforcement, and lending. Without intentional safeguards, AI risks deepening inequality and eroding public trust.

We must recognize that AI has not emerged in a cultural vacuum. It is vital to remember that AI is shaped by the intentions and values of its designers and users. And the values shaping its development differ across regions—Europe's emphasis on privacy, America's drive for innovation, or China's focus on social control—each raising distinct challenges for the global Church. From a theological perspective, the Church must consider how AI can be developed and deployed in ways that promote justice, compassion, and human flourishing. It has a vital role to play in shaping this conversation, advocating for policies and practices that reflect the values of the Kingdom of God while challenging technologies that exploit, manipulate, or dehumanize.

Just as the printing press transformed the dissemination of the Gospel in the fifteenth century, AI now holds the potential to reshape how the Church engages with the world. Whether it becomes an instrument of grace and peace or a tool of distortion and harm will depend on how it is stewarded—and how we care for and form those who create it.

One of the most basic ethical concerns is accountability: if AI causes harm, who is responsible—the user, the programmer, or the system's owner? Without transparency in how AI makes decisions, accountability remains elusive, and here the Church has a crucial role in calling for justice and fairness.[11]

Why AI Matters for the Church

For church leaders and Christian communities, AI functions as both a powerful instrument and a profound challenge. It offers opportunities for ministry innovation, efficiency, and outreach, while also introducing questions that touch on ethics, human dignity, and theology. And the Church cannot simply remain a passive observer as these changes unfold. Understanding what AI is—and what it is not—is essential for guiding congregations through the spiritual and cultural transformations that accompany such significant technological advancements.

Looking ahead, AI will continue to evolve, raising new and complex questions for society in general and particularly for the Church. Could machines ever be considered moral agents? How should the Church respond if AI begins to replicate aspects of pastoral care? What does authentic human connection look like in an increasingly automated society? Church leaders may not need to master programming, but they must become conversant in the language and implications of AI. Such literacy will equip them to care for their congregations, engage meaningfully in broader cultural debates, and ensure that technology remains a servant of the Gospel rather than a substitute for it.

Chapter 1:
Questions for Reflection

1. The chapter highlights that, while AI is often mistaken for "thinking," in reality, it is pattern recognition without comprehension. How does that distinction shape your trust—or caution—about using AI in ministry?

2. From Talos in Greek mythology to Alan Turing in the modern era, the human desire to create artificial life has its roots in antiquity. What does this continuity suggest about human creativity and our longing to imitate God as Creator?

3. You read that AI mirrors the values and biases of its creators. How might this truth inform the way the Church approaches questions of justice, equity, and discernment in technology?

4. The chapter compares AI's impact to the printing press in reshaping Christian witness. What parallels and differences do you see between those two moments in history?

5. Accountability remains a central ethical challenge: if AI causes harm, who should be held responsible? How might Christian theology of sin, stewardship, and justice guide our answer to this question?

Chapter 2:
How AI Works in Everyday Life

The way to get artificial intelligence really smart is to get it to learn from everyday life.

~ Geoffrey Hinton
pioneer of deep learning[12]

An Invisible Companion

Artificial Intelligence is no longer a novelty—it is an unseen companion that works silently behind the scenes of modern life. From the moment much of the world's population wakes up, checks their phones, and scrolls through their feeds, AI has already begun shaping their day. It influences the messages we see, the routes we take to work, the way we shop, and even the healthcare we receive. While its presence is often invisible, AI is deeply woven into the fabric of daily routines, making it a constant but usually unacknowledged force in the world.[13]

Shaping Our Digital Lives

One of the most familiar encounters with AI comes through personal digital assistants like Apple's Siri, Google Assistant, and Amazon's Alexa. These voice-activated tools rely on sophisticated natural language processing to understand questions, retrieve information, and carry out commands. A simple request, such as to play a song, set a reminder for a meeting, or find a nearby coffee shop, activates multiple layers of AI technology—speech recognition, language interpretation, contextual analysis, and information retrieval. Even the predictive text and autocorrect functions on smartphones are powered by AI models trained to learn typing patterns, anticipate the next word, and improve accuracy over time. The same underlying technology drives translation services like Google Translate, which can instantly convert text or speech into another language, enabling real-time communication across cultures and continents.[14]

Another space where AI exerts considerable influence is social media platforms. Facebook, Instagram, X, TikTok, and other platforms all deploy AI-driven algorithms that analyze user interactions, likes, and viewing habits to personalize the content presented in their feeds. The videos and posts a person sees are not random but curated by vast networks of machine learning models that predict which content will capture attention. At the same time, AI can also work in the background to detect and remove harmful content, combat various levels of misinformation, and flag hate speech or cyberbullying. These moderation systems, while imperfect, represent a growing reliance on automated decision-making to maintain safe digital spaces.[15]

Underlying all these systems is the collection of massive amounts of personal data from users. Every click, view, and interaction is tracked and analyzed to improve predictions. While this enables convenience and personalization, it also raises concerns about surveillance, privacy, and the ways people's choices can be subtly shaped by algorithms rather than being made freely.[16]

Search engines like Google and Bing also depend heavily on AI to deliver relevant and accurate results to user requests. Every search query is analyzed for intent and context, then matched against countless indexed web pages using ranking algorithms that evaluate keyword relevance, site authority, and user engagement patterns. Increasingly, searches are not limited to typed queries—voice search enables people to speak their requests to their devices, with AI parsing and understanding their words to produce relevant results. This shift has transformed how people access information, moving from a world of static search results to dynamic, context-aware answers.[17] These same systems are being refined to identify credible sources and alert users when they may encounter unreliable or misleading information.

Personalization and Power

In online shopping, AI transforms the consumer experience from simple browsing to personalized engagement. When someone shops on Amazon, eBay, Shopify, or a wide range of other available online platforms, AI recommendation engines draw on the person's browsing history, past purchases, and general demographic data to suggest products tailored to individual preferences.[18] This personalization extends far beyond marketing— AI models help the seller predict inventory needs, optimize warehouse operations, and streamline delivery logistics, ensuring products are available when and where they are needed.

In many cases, these systems operate without the consumer ever realizing that AI is facilitating the process. As these systems become more effective, the line between human choice and machine suggestion begins to blur. What feels like a personal decision may in fact be a response to algorithmic nudges carefully optimized to capture attention or drive consumption.

Just as AI personalizes and streamlines online shopping, its influence extends with equal force into the financial world, where speed, accuracy, and protection are paramount. Automated fraud detection systems analyze transaction patterns in real time, identifying suspicious activity before damage can occur. AI-driven credit scoring utilizes a broader range of data—such as spending habits, employment stability, and payment history—to assess creditworthiness with greater speed and precision than traditional models. Meanwhile, robo-advisors utilize predictive analytics to provide personalized investment advice, enabling individuals to manage their portfolios with strategies informed by vast amounts of financial data.

Yet these same systems can also reflect and reinforce bias. Credit models may disadvantage certain groups, medical algorithms may misdiagnose based on unrepresentative data, and predictive policing systems have been shown to target minority communities disproportionately.[19] In each case, the promise of efficiency collides with the desire for justice and fairness.

Transforming Health, Mobility, and Home

Beyond finance, AI's ability to analyze complex data and recognize patterns is also proving as transformative in healthcare, where the stakes involve not money and merchandise but human lives. Machine learning models can examine medical records, genetic data, and imaging scans to detect early signs of disease and support accurate diagnoses. Radiology departments increasingly use AI-enhanced imaging tools to identify abnormalities in X-rays, MRIs, and CT scans, often spotting patterns that human eyes might overlook.[20] In parallel, AI-powered chatbots and virtual assistants now provide patients with basic medical guidance, directing them to appropriate resources or specialists and offering support between in-person visits.

Just as AI is redefining healthcare, it is also steering significant changes in transportation, where the desire for greater efficiency and safety drive innovation. Navigation platforms, such as Google Maps and Waze, process real-time traffic conditions, weather data, and road closures to recommend the most optimal routes for motorists. Ride-hailing services like Uber and Lyft depend on AI to match passengers with drivers, minimize wait times, and predict areas of high demand. Even more transformative is the ongoing development of self-driving car technology, where AI controls steering, acceleration, braking, and environmental awareness to improve safety and efficiency on the roads.[21]

While AI is transforming how we move through the world, it is also reshaping the spaces we return to, embedding intelligence into the very fabric of our homes. Smart thermostats, like Honeywell, Ecobee and Nest, learn household routines and automatically adjust temperatures to optimize efficiency and comfort. Security systems, such as Eufy, Arlo and Ring analyze motion data to distinguish between a passing car and a person approaching the door. Lighting systems respond to both user preferences and environmental cues, creating adaptive environments that make homes more responsive and efficient. These devices, among many others, are part of the Internet of Things (IOT)—a network of AI-enhanced appliances and sensors that communicate with each other to simplify daily life.[22]

Entertainment, Education, and the Hidden Costs

AI also shapes how we relax and engage with culture, driving a new era of personalized and even AI-generated entertainment. Streaming services like Netflix, Spotify, and YouTube use sophisticated algorithms to recommend content based on previous listening or viewing patterns. These systems continuously refine their suggestions to keep users engaged. AI has even demonstrated the capability to create entertainment, generating music compositions, writing scripts, and producing visual art. The rise of what is known as deepfake technology—AI-generated synthetic media that convincingly alters or fabricates audio and video—demonstrates both the creative potential and the ethical challenges of AI-driven media production.[23]

Just as AI entertains and challenges us with personalized media, it is also impacting how we learn, tailoring education to individual needs and opening new pathways to knowledge. Platforms such as Duolingo, Coursera, and Khan Academy utilize AI to tailor lessons to each student's pace and level of understanding. Virtual tutors respond to questions instantly, while AI-driven feedback systems help learners correct mistakes in real time.[24] This individualized approach makes learning more interactive and accessible, breaking down barriers of geography and cost.

Still, despite its great potential, the benefits of AI are not evenly distributed, and its impact is not without its drawbacks. Large portions of the world lack the infrastructure or resources to access these technological tools, widening the digital divide between the connected and the disconnected. And beneath the convenience and efficiency of this everyday AI also lies the hidden cost of energy. Training and running large AI models requires enormous computing power, drawing on the power grid and contributing to carbon emissions and environmental strain. Even the simple act of streaming music or running a translation service relies on vast data centers that consume significant amounts of electricity and water. For the global Church, these issues raise questions of how technology can be shared in ways that promote equity rather than deepen inequality and how we can do so while still caring for the Creation that we've been entrusted to steward.[25]

Beyond the classroom, AI's capacity to analyze data and optimize processes is equally transformative in industry and government—often working quietly behind the scenes. Businesses harness AI-driven analytics to forecast demand, allocate resources efficiently, and anticipate equipment failures before they occur. Governments employ similar systems to interpret public data, strengthen infrastructure, and enhance public safety. In these spheres, AI's invisible efficiency shapes daily life in ways most people rarely notice.

Yet this very pervasiveness brings new challenges. As we grow accustomed to AI's reliability and speed, we may be tempted to treat its judgments as authoritative, trusting algorithms more than our own discernment. The same systems that promise insight and order also confront us with questions about privacy, algorithmic fairness, and the future of human work—inviting deeper reflection on how technological progress should serve human dignity and the common good.[26]

The integration of AI into everyday life is so seamless that many people fail to notice it. It shapes digital interactions, personalizes services, and drives decisions across countless sectors. It is as present in the seemingly trivial task of suggesting a new song as it is in the life-or-death decision-making of medical diagnosis. While the benefits of this technology are vast, so too are the ethical responsibilities that accompany its use. AI's growing role in society demands not only technical literacy but also moral discernment, ensuring that its power is directed toward the good of individuals, communities, and creation itself.

Chapter 2:
Questions for Reflection

1. Much of AI's work happens invisibly in the background of daily routines. How does this hidden presence affect the way you think about dependence on technology?

2. Predictive algorithms often personalize what we see, buy, or learn. In your own life, when do you feel most aware—and least aware—that AI suggestions are shaping your choices?

3. This chapter raises concerns about surveillance, privacy, bias, and fairness in a variety of AI systems. Which of these concerns feels most pressing for your community and why?

4. Online platforms increasingly filter content through personalized feeds and recommendations. What habits or practices help you stay rooted in Christ when algorithms are subtly shaping your attention?

5. The chapter ends by noting that people may begin trusting algorithmic decisions more than human discernment. How should the Church respond when efficiency tempts us to surrender wisdom to machines?

Chapter 3:
The AI Revolution and the Church

The Church has to become increasingly intelligent about technology, because technology is increasingly about intelligence.

~ N. T. Wright,
New Testament scholar
and theologian[27]

Learning from Past Revolutions

Throughout history, the Church has lived through and responded to waves of technological transformation that altered the way humanity works, communicates, and organizes society. The invention of the printing press in the fifteenth century democratized access to Scripture, allowing ordinary believers to read the Bible for themselves and empowering movements like the Protestant Reformation.[28] In the industrial era, mechanization transformed economies by allowing for the development of large-scale manufacturing. This change, like the automation revolution that followed and the contemporary 24/7 business model that followed, uprooted traditional ways of life, prompting the Church to speak out on issues such as worker exploitation, fair wages, and the observance of the Sabbath.[29]

In the digital age, the rapid rise of television, radio, and later the internet reshaped communication and community formation, leading many churches to adopt broadcasting and online platforms to reach global audiences.[30] Each of these technological revolutions required discernment, adaptability, and a renewed theological vision for ministry in a changing world.

What Makes AI Different

Today, artificial intelligence represents another such turning point, but one with unprecedented reach. Unlike earlier technologies that primarily extended human physical capacity or sped up information sharing, AI introduces the ability to replicate—and in some cases surpass—human decision-making, problem-solving, and creative expression. It does not simply make work more efficient; it can, in specific contexts, perform work on our behalf. This includes tasks once thought to be uniquely human, such as drafting legal documents, composing music, diagnosing medical conditions, and managing complex logistics.[31] In this sense, AI carries the potential to disrupt both manual labor and white-collar professions, impacting employment across the entire socioeconomic spectrum.

The economic implications of this change are profound. In industries built on repetitive, predictable tasks—such as manufacturing, data entry or analysis—AI systems are already beginning to replace humans. Automation of this sort can lead to significant gains in productivity and lower costs for businesses and governments, but it can also result in massive-scale job displacement and unemployment. The transition to a more AI-driven economy may deepen income inequality if new opportunities disproportionately benefit those with advanced technical skills while leaving behind those whose work is more easily automated. Unlike previous industrial shifts, the pace of AI adoption is moving so quickly it is compressing decades of change into a few short years.[32]

And it is also important to note that the effects of AI will not be uniform globally. Churches in wealthier nations may struggle with the impact of job automation and digital abundance, while those in less-resourced contexts may face unequal access to these technologies, thereby widening the digital divide.[33] This will undoubtedly create a range of challenges that the Church will need to consider and address, with widespread systemic implications.

Work, Vocation, and Human Dignity

This new reality will challenge global understandings of work, purpose, and vocation. For centuries, Christian theology has affirmed that work is not merely a means of survival but a calling—a way to participate in God's ongoing creation and care for the world. When AI takes over tasks that once formed the foundation of people's livelihoods, it raises pressing questions: How will displaced workers find new ways to exercise their gifts? What happens to communities whose economic stability depends on industries vulnerable to automation? How should believers understand their identity and purpose when traditional pathways of work are altered or removed?

The Church's Call In the AI Age

Here, the Church is uniquely positioned to stand in the gap as both a prophetic voice and a pastoral presence. Theologically, we as church leaders are called to reimagine work not merely as economic activity but as a vocation that contributes to the common good—whether expressed through caregiving, mentoring, volunteering, or creative endeavor.[34] Practically, this vision takes shape as congregations become centers of formation, equipping people for lifelong learning and resilience in the face of technological disruption. Through digital literacy workshops, vocational discernment groups, and connections to community reskilling initiatives, the Church can help its members adapt faithfully to change. Spiritually, it offers a deeper word of hope: that human worth is never reducible to productivity but is grounded in the unshakable truth of being beloved children of God.

Equally important is spiritual formation. Because AI systems increasingly shape what we see, hear, and even desire, the Church must help disciples cultivate practices of discernment, prayer, and imagination that resist being passively formed by algorithms. In doing so, the Church guards against technology discipling believers more than the Gospel does.

Additionally, AI also raises complex ethical questions that demand the Church's engagement. Decisions once made by human beings—such as whether to grant a loan, hire an employee, or approve a medical treatment—are, either in part or wholly, increasingly being delegated to algorithms. These systems can inherit and even amplify biases present in the data they are trained on, leading to unjust or unfair outcomes that may disproportionately harm marginalized communities.[35]

Moreover, the vast collection and analysis of personal data required to power many AI systems introduces serious concerns about privacy and individual autonomy and control over their personal information.[36] If left unchecked, these developments could erode trust, diminish human agency, and concentrate power in the hands of a small number of corporations or governments.

And all of this points to the fact that, when it comes to technology, the Church has a crucial role to play as an advocate for those who are most vulnerable and an ethical voice to those who make the critical development and deployment decisions. It will need to speak up to make sure that AI and other tools are developed in ways that protect human dignity and promote justice. This might involve participating in public policy conversations, supporting regulations that ensure AI remains transparent and accountable, or even coming together in mass to encourage tech companies to adhere to ethical guidelines based on fairness, compassion, and respect for all.[37] Just as importantly, the Church can help its members understand the moral questions raised by AI, enabling them to respond wisely and prayerfully in their jobs, families, and communities.

The AI revolution also presents extraordinary opportunities for ministry. Just as the printing press enabled the mass distribution of Scripture, AI can facilitate the translation of the Bible into languages that have never had a written text, breaking down barriers to the Gospel.[38] It can help churches reach dispersed congregations through automated transcription, translation, and online content delivery. It can assist pastors with sermon preparation, research, and administrative tasks, thereby freeing up time for deeper, relational ministry.

Still, with every new possibility comes a responsibility to use it wisely, which means that these opportunities ought to be pursued with discernment, ensuring that technological efficiency doesn't fully replace the personal presence and pastoral care that lie at the heart of Christian community. There is also the subtle danger of idolatry. When AI systems become so trusted that they are treated as ultimate authorities, they risk taking a place that belongs only to God.[39] The Church must remind people that while technology can be a powerful servant, it makes a poor master.

Shepherding Through Uncertainty

And that is because, ultimately, the Church's role in the AI era is to shepherd people through uncertainty, anchor them in the truth of God, and model a faithful engagement with technology. History shows that when the Church approaches technological shifts with both courage and humility, it can help shape those changes for the good of society.[40] The AI revolution will test that ability anew. By drawing on its deep well of theological reflection, ecumenical and interfaith collaboration, its commitment to justice, and its tradition of adapting to new tools for the sake of the Gospel, the Church can guide communities through this transformative age—ensuring that AI serves humanity rather than the other way around.

Chapter 3:
Questions for Reflection

1. The Church has responded to many waves of technological disruption. What patterns of wisdom or mistakes from those responses could guide the way we navigate AI today?

2. How has your own understanding of vocation been challenged or deepened by experiences with technologies—and how might AI continue that shift?

3. When algorithms compete for your attention, what practices specifically help keep your identity rooted in Christ rather than in digital influence?

4. AI raises concerns about bias, privacy, and the concentration of power. In what ways can you practically imagine being an advocate for the vulnerable or a moral guide for those shaping technology—to help ensure AI serves human dignity and the common good?

5. The chapter warns that AI can become an idol when treated as an ultimate authority. How can the Church remind people that technology should remain a servant, not a master?

Part 2:
AI in Ministry

Chapter 4:
How AI Becomes Our Digital Mirror

Technology is a mirror. It shows us who we are, not who we wish to be.

~ Sherry Turkle sociologist and author of Alone Together[41]

AI as a Mirror of Humanity

Artificial Intelligence is more than a neutral tool; it is a mirror that reflects our values, biases, fears, and aspirations back to us in digital form.[42] It learns from our collective history—our art, our language, our social patterns—and in doing so, it reveals both the beauty and the brokenness of the human condition. As AI continues to advance, it forces us to wrestle with profound questions about human worth, especially as machines begin to mimic intelligence, creativity, and even emotional connection. This reflection compels us to examine our identity, relationships, and spiritual formation in ways that previous technologies never demanded.

AI learns from us. It is currently primarily shaped by the data we feed it, data drawn from the vast archives of human experience. Every photograph, article, video, and conversation that becomes part of its training material is a fragment of who we are as a global society. The algorithms that power social media platforms, content recommendation systems, and conversational agents are reflective surfaces, absorbing our biases alongside our aspirations.

Yet it is crucial to remember that this mirror is not neutral. AI reflects the cultures and languages most represented in its data, which often means Western and English-speaking perspectives dominate.[43] As a result, voices from smaller cultures and marginalized communities may be distorted, silenced, or misrepresented. When such systems perpetuate discrimination, they do not create new prejudice but rather magnify the injustices embedded in their training data. Conversely, when designed intentionally and ethically, AI can be used to identify inequities and inspire redemptive change.

Creativity and the Image of God

One of the most striking ways AI holds up this mirror is in the realm of creativity. Today, algorithms can compose symphonies, paint portraits, write poetry, and generate vivid cinematic scenes.[44] These capabilities press us to ask whether creativity itself is an exclusively organic gift or whether machines can meaningfully participate in acts of creation.

For centuries, Christian thought has linked creativity to the *imago Dei*—the image of God in humanity—understanding our creative capacity as a reflection of the Creator.[45] As AI continues to improve in imitating human artistic expression with increasing authenticity and even superiority, we will have to ask whether our worth lies in our output or in something more profound, such as the intention, relationship, and spiritual reality behind our actions.

Relationships in a Digital Age

The reflection in the mirror becomes even more intimate when AI enters the realm of relationships. Digital companions, from chatbots to robotic caregivers, are increasingly being used to provide comfort and support, articularly to the elderly, the isolated, and those who live with chronic loneliness.[46] For some, these aids offer genuine relief and a sense of connection; yet for others, they raise concerns about authenticity and dependence. Either way, AI invites us to ponder what it means to be truly known and loved, not just remembered and responded to. In this tension, we are reminded of the gift of presence and the sacred depth of relationships that we've been given.

Additionally, platforms use social algorithms to curate our online worlds, filtering what we see and shaping our sense of reality. Personalized content feeds may reinforce our existing beliefs, creating ideological echo chambers that insulate us from differing perspectives.[47] While this personalization can feel affirming, it can also subtly distort our worldview.

The question is whether AI helps us form deeper connections in ways once unimaginable, or whether it quietly erodes them by replacing authentic, face-to-face community with curated, self-reinforcing digital experiences. This latter possibility potentially threatens the communal nature of Christian faith. When discipleship is reduced to individualized content streams, the Church risks losing the shared practices of gathering, worshiping, and discerning together as one body.

Spiritual Formation and Commodification

In the realm of faith, the role of AI in spiritual formation is rapidly expanding. Apps now offer AI-generated prayers, guided Bible study plans, and even experimental tools that assist in creating sermons, which can be customized to individual preferences.[48] Virtual "clergy" and pastoral chatbots provide quick answers to theological questions, drawing on massive databases of Scripture and commentary.

While these tools can make faith resources more accessible, they also risk reducing spiritual life to a consumer-driven experience, where believers receive algorithmically optimized content rather than engaging in the relational and transformative work of discipleship in community with others. Can a prayer generated by a machine carry the same spiritual weight as one uttered from a human heart? Can a sermon designed to please a listener's preferences still challenge, convict, and call to repentance?

This reality underscores the need for spiritual disciplines—such as prayer, Scripture meditation, Sabbath observance, and communal worship—as counterweights to the curated, convenience-driven spirituality that AI often encourages.[49] The Church must not only evaluate the technologies but also actively form believers who can engage with them without losing sight of the practices that anchor faith in God, rather than in persuasive algorithms.

Another concern is that many of the technological tools are produced by commercial enterprises with priorities that may not align with the Church's mission. When worship and discipleship become mediated by platforms driven by profit, there is a risk that faith itself becomes commodified, shaped more by market forces than by the Spirit.[50]

Theological Reckoning and Christian Witness

Moreover, AI's prevalence forces a larger theological reckoning: if AI can simulate emotional intelligence, produce art, and offer what appears to be pastoral care, does it redefine what it means to be created in the image of God? Historically, theologians have associated the *imago Dei* with traits such as rationality, creativity, and relational capacity.[51] As AI demonstrates imitations, improvements, and recreations of these traits, the Church must clarify which aspects of humanity are essential and inimitable.

Looking forward, AI will continue to prompt us to engage in deeper reflection on these matters. The Church has a unique responsibility to guide this reflection, ensuring that technology serves human flourishing rather than diminishing human dignity. As we stand at this threshold, the challenge is not simply to adopt new tools, but to discern how they can deepen our love of God and our neighbor. The Church's witness will likely be measured not by its technological sophistication but by its ability to embody Christ's presence in a world increasingly mediated by machines.

If AI is indeed a digital mirror, then the question is not whether it will reflect us, but what version of ourselves we will see when we look into it. Will we see an image distorted by our self-interest, prejudice, and fear or by our generosity and grace and hope? Will we use the reflection to be affirmed in our comfort or as an opportunity to pursue greater integrity, compassion, and faithfulness? The objective measure of our humanity in the age of AI will not be our ability to outthink or out-create machines, but our commitment to embody love, justice, and truth in a world increasingly mediated by technology.

Ultimately, our worth will not be determined by how closely we can compete with the capabilities of AI, but by how faithfully we live into the call to love God and our neighbor. The mirror AI holds before us can either lead to vanity and self-deception or to repentance and renewal. The choice, and the responsibility, remain ours.

Chapter 4:
Questions for Reflection

1. If AI is a mirror of humanity, what parts of our culture and character do you most see reflected in it today?

2. AI-generated art, music, and writing raise new questions about creativity. Do you think creativity is only valuable when it flows from human intention and spirit, or can machine-made works still reveal something of God's gift?

3. Digital companions and social algorithms are changing how people form relationships. What do you think is lost—or gained—when connection is mediated primarily through machines?

4. When faith practices are shaped by algorithmically generated prayers, sermons, or recommendations, how can the Church ensure that discipleship remains relational and Spirit-led rather than commodified?

5. The chapter suggests that AI forces us to clarify what it means to be made in God's image. Which aspects of humanity do you believe are essential and inimitable, no matter how advanced AI becomes?

Chapter 5:
AI in Worship, Discipleship & Missions

Technology as a Servant in Ministry

Artificial intelligence is already reshaping how the Church worships, disciples believers, and engages in missions.[53] No longer confined to the realm of technological novelty, AI is becoming a practical ministry tool—enhancing theological research, deepening Bible study, streamlining communication, and expanding the global reach of the Gospel. This transformation is not simply the result of increased efficiency in performing routine tasks; it is about harnessing the technology to actively serve the purposes of God's Kingdom. The challenge for church leaders is to do this while preserving the authenticity, relational depth, and Spirit-led character of Christian ministry.

Worship in a Digital Age

As churches begin to explore these new possibilities, some of the most immediate and visible areas of impact are in worship. Worship leaders and pastors now have access to AI-powered tools that can assist in planning services, composing music, drafting sermons, and engaging congregations in new ways. These tools can help congregations craft worship sets that are thematically tied to a sermon series or the liturgical calendar.

AI can also assist pastors in sermon preparation by quickly gathering relevant Scripture references, consulting commentaries, and suggesting illustrative material. It can generate initial outlines that a preacher can then refine, ensuring more time is spent in prayer and reflection rather than on administrative searching.[55] Algorithms are capable of analyzing vast libraries of worship songs, identifying lyrical and thematic patterns, and even generating new compositions that align with a church's theological convictions.[56] Beyond in-person services, AI-powered livestream enhancements and alternative virtual worship environments enable believers to participate from anywhere in the world, creating digital sanctuaries for those who are homebound or traveling. This can also make it possible for those living in regions without access to local churches to participate.[57]

Discipleship and Spiritual Growth

Just as worship reflects the Church's shared heartbeat, discipleship expresses its ongoing walk of faith. And here too, AI is beginning to shape the journey—helping believers grow in wisdom, deepen their practice, and encounter God's Word in profoundly personal ways. AI-driven discipleship platforms can adapt Bible study plans to individual learning styles, recommend Scripture readings specific to a believer's current struggles or questions, and provide cross-referenced theological insights in multiple languages.[58]

Chatbot-based guides offer real-time responses to theological questions, suggest devotional readings, and provide written prayers in response to personal requests. These systems can track spiritual practices, such as prayer frequency, Scripture memorization, and acts of service, offering encouragement and new challenges tailored to the believer's journey.[59] AI can also help small group leaders by suggesting contextually relevant discussion questions, identifying biblical passages that connect to current events, and providing resources that make gatherings more engaging and interactive.

Missions, Evangelism, and Global Reach

Just as discipleship equips believers to grow in faith and maturity, missions turns that growth outward in service to the world. Here too, AI is extending the Church's reach—opening doors for evangelism, translation, and compassionate action that were once unthinkable.

Digital evangelism campaigns can now utilize AI-driven analytics to understand audience engagement, test various messages, and pinpoint the most effective ways to share the Gospel with diverse audiences. Social media targeting enables ministries to connect with individuals who have expressed interest in spiritual topics, delivering content in culturally and linguistically resonant ways.[60] Translation tools, powered by machine learning, have enabled the communication of Scripture and faith materials in hundreds of languages, thereby breaking down barriers that previously limited missionary effectiveness.[61]

Yet even as these tools expand access, they risk carrying cultural and theological assumptions from the contexts in which they were built rather than those to which they are directed. Evangelism is best when it strives to be contextual and sensitive to local cultures and traditions of the intended audience. We must guard against allowing our technology to inadvertently export cultural bias rather than the Good News.

Beyond evangelism, AI supports humanitarian and development work by helping mission organizations anticipate areas of need, optimize supply chains for aid delivery, and identify vulnerable communities before crises occur. In urban settings, AI can analyze demographic trends, religious engagement patterns, and social challenges to guide strategic church planting in areas most in need of a Gospel witness.[62]

Risks, Discernment, and the Path Ahead

While these innovations expand the Church's global reach in remarkable ways, they also introduce new complexities. The same algorithms that help translate Scripture into hundreds of languages can unintentionally carry cultural bias; the same analytics that guide evangelism can shape theology through unseen assumptions. As the Church embraces these tools for mission and ministry, discernment becomes not just advisable but essential.

In worship, present-day AI-generated sermons and music compositions may lack the lived spiritual depth that comes from human experience and divine inspiration. There is a risk that congregations could become passive consumers of polished but impersonal content rather than active participants in Spirit-led worship.[63]

Additionally, many communities lack the necessary infrastructure and access to benefit from these technologies. Without attention to equity, the AI revolution in missions could deepen the digital divide.[64] The result of this would be privileging already-connected churches while leaving others further behind.

In discipleship, overreliance on AI tools could weaken the relational and communal aspects of spiritual formation, replacing face-to-face mentorship with convenient but ultimately shallow digital interactions. Privacy protection also emerges as a significant concern, since AI-driven platforms often depend on large amounts of personal data to function effectively.[65] This reliance, in turn, raises ethical questions around data security, informed consent, and the potential for misuse.

Beyond these risks, the fact that AI systems are trained on existing datasets means they may also carry the cultural or theological biases embedded in that data, subtly shaping the content they deliver and influencing how faith and practice are understood.[66] This raises another crucial question. If sermons, devotionals, or Bible study guides are entirely generated by AI, who is the teacher—the pastor, the algorithm, or the unseen programmers who shaped it?

Despite these concerns, the potential of AI to enhance ministry is obviously enormous—if it is approached with wisdom. The Church's task is to ensure that AI serves as a servant, not a master; a tool that complements the Spirit's work rather than replacing it. This will mean setting boundaries, maintaining appropriate oversight in areas such as teaching and pastoral care, and integrating AI in ways that promote participation rather than passive consumption. The role of church leaders is to assure that theological integrity, relational depth, and a clear sense of mission remain at the center of all technological adoption.

And as we look ahead, we can expect AI's role in worship, discipleship, and missions to expand even further. We may see more sophisticated sermon-writing assistants that draw from global theological traditions, AI-generated devotionals tailored to progressing spiritual seasons of individual believers, and immersive extended reality church services that bring global congregations together in shared worship. AI-powered virtual missionaries could one day engage in initial Gospel conversations in places where human missionaries cannot safely go, preparing the way for in-person follow-up.[67] The possibilities are vast—but so is the need for prayerful discernment.

In the end, AI's value to the Church will be determined not by its technical brilliance but, again, by how it is used to glorify God and build God's Kingdom. When deployed with humility, guided by Scripture, and anchored in love, AI can help the Church worship more deeply, disciple more effectively, and reach the nations with renewed creativity. The challenge before us is to harness this technology in a way that strengthens, rather than supplants, the human and divine relationships at the core of the Christian faith.

Chapter 5:
Questions for Reflection

1. This chapter describes how AI can compose music, prepare sermon outlines, and even generate devotionals. How can the Church distinguish when technology enhances worship and when it risks hollowing out spiritual depth?

2. Discipleship is often understood as a relational and communal experience, yet AI offers individualized and personalized faith guidance. How might your church balance the use of digital tools with the need for human mentors and embodied fellowship?

3. AI translation and targeting tools allow the Gospel to reach new audiences in creative ways. How can churches ensure that these tools communicate the Good News faithfully, without exporting unintended cultural or theological bias?

4. AI-powered missions and humanitarian tools can predict needs and deliver aid with efficiency. What safeguards should be in place to make sure efficiency doesn't eclipse compassion and local wisdom?

5. The chapter raises the concern that sermons and lessons generated by AI may carry the biases of unseen programmers. Who do you believe should bear ultimate responsibility for teaching and shaping discipleship in an AI-driven church?

Chapter 6:
AI in Stewardship and Care

The Church is the only society that exists for the benefit of those who are not its members.

~ William Temple
Archbishop of Canterbury[68]

Tools for Healthy Ministry Systems

Artificial intelligence is emerging as one of the Church's most versatile tools for both operational stewardship and pastoral care.[69] By streamlining communication, optimizing scheduling, and enhancing financial management, AI enables ministry leaders to complete routine tasks with greater efficiency, freeing them to focus on spiritual leadership and relational ministry. Additionally, beyond the administrative sphere, AI is increasingly being integrated into donor engagement, mental health support, counseling assistance, and personalized discipleship—creating opportunities to serve people more effectively while challenging the Church to maintain its ethical and spiritual commitments.

Healthy ministry begins with healthy systems. Churches, whether small congregations or large multi-campus ministries, depend on well-organized administration to function smoothly, grow, and maintain holistic health. AI-powered tools are now capable of handling many of the time-consuming tasks that once preoccupied staff hours.

Intelligent scheduling assistants can automatically coordinate pastoral appointments, reserve facilities, and arrange volunteer shifts, reducing scheduling conflicts and last-minute changes. AI-enabled chatbots can provide instant answers to frequently asked questions—everything from service times and event details to childcare availability—allowing members and visitors to find information quickly without waiting for a human reply. Automated communication platforms can segment audiences and deliver targeted messages through email, text, or social media, ensuring that reminders, devotionals, or event invitations reach the right people at the right time.[70]

Financial Stewardship and Donor Engagement

Just as AI streamlines the administrative systems that support the ministry, it also strengthens the financial stewardship on which those systems depend. AI-powered financial systems can track donations in real time, flag irregularities, and project giving trends based on historical data.[71] This allows church leaders to make informed budgeting decisions, and can help ensure that funds are allocated wisely and transparently. AI does this through data-driven budgeting, predictive modeling, automatically categorizing spending, flagging anomalies, and creating real-time reports which reduce human error and bias in allocation helping to align spending with stated values and strategic goals.

Donor management platforms, enhanced by AI, can send personalized thank-you notes, create customized giving reports, and suggest opportunities for involvement that align with a supporter's interests and previous contributions.[72] By automating these interactions, churches can sustain strong donor relationships without overburdening staff. Importantly, AI-driven analytics can also help identify areas where stewardship practices can be improved, promoting a culture of accountability, trust, and generosity within the congregation.

AI in Pastoral Care and Discipleship

The role of AI in pastoral care is perhaps the most sensitive and potentially transformative use of this technology. In a world where mental health needs are being identified more readily, AI-driven chatbots and virtual counseling assistants can serve as a first line of contact for those seeking support.[74] These systems can offer Scripture-based encouragement, provide resources on coping strategies, and guide individuals toward human counselors when deeper engagement is needed. Available 24/7, AI can ensure that no one in distress is ever left entirely without support.

Sentiment analysis tools, which aim to detect emotional cues in written or spoken communication, are being explored as aids for church leaders—offering potential signals when a congregant's language may suggest loneliness, grief, or crisis.[75] These tools are not providing definitive diagnoses, but they can serve as prompts for deeper pastoral attention and care. By monitoring trends in prayer requests, survey responses, or social media interactions, church leaders can proactively reach out to those in need before a problem escalates.

Personalized discipleship is an emerging area where AI shows promise in strengthening pastoral care. A growing number of apps already offer adaptive Bible study paths, customized devotionals, or suggested content based on a person's stated interests or prior engagement.[76] For example, someone drawn to themes of forgiveness might be guided toward resources on reconciliation and peacemaking. Prayer reminder systems and Scripture memorization tools are also beginning to incorporate personalization, adjusting to individual participant's schedules and learning patterns. While these tools remain limited in scope, they are gradually providing pastors with new insights that can support mentoring, small group development, and congregational teaching priorities.

These possibilities also highlight the need to form church leaders who can use AI wisely. Without theological training and ethical grounding, leaders may over-rely on technology or misuse its insights. Preparing pastors and lay leaders for digital discernment is itself part of faithful stewardship.[77]

Boundaries, Pastoral Ethics, and Transparency

As noted previously, as promising as these technological developments are, they also come with significant ethical considerations and challenges. As church leaders consider the use of AI in ministry, it is hoped that Christian values and a deep respect for human dignity will shape the application of whatever tools are deployed. Because, while automation can enhance ministry, it shouldn't replace the relational heart of the Church's mission.

In counseling, for instance, AI can offer helpful guidance but cannot currently replicate the empathy, discernment, and Spirit-led wisdom of a church leader. Additionally, faithful stewardship also requires naming boundaries. Some functions of ministry—such as administering sacraments, hearing confessions, or offering Spirit-led counsel— may belong uniquely to human shepherds.[78] And likewise, we may find that some functions belong uniquely to AI shepherds.

Either way, these distinctions make transparency essential—especially when technology is applied in sensitive areas such as mental health or spiritual care. This means being open and forthright about how and when AI is being used, what kind of data it collects or analyzes, and how decisions derived from that data are made. It involves informing congregants when digital tools are shaping pastoral care, counseling interactions, or ministry recommendations— especially in contexts that touch on personal, emotional, or spiritual well-being.

And with transparency comes another pressing concern: data privacy.[79] Congregants need assurance that their personal information will be safeguarded, not misused or mishandled. Because AI systems learn from existing datasets, they can unintentionally replicate biases present in the data, which can subtly influence pastoral recommendations or outreach strategies.

These concerns remind us that while AI offers new possibilities, its role in ministry must always be carefully discerned.

Equally important is remembering that data points never tell the whole story of a person. A dashboard may flag loneliness or crisis, but proper care and appropriate intervention comes from seeing congregants as children of God, not merely as patterns in a dataset.[80] Ethical vigilance, transparency, and accountability provide the guardrails—but discernment also points toward the deeper question of how AI can be used faithfully.

Toward Faithful Stewardship

Generally speaking, the key to using AI in the stewardship of the Church and its mission is balance. When used wisely, AI can reduce the time spent on administrative tasks, allowing ministry leaders to devote more time to prayer, teaching, and personal relationships. It can expand the Church's capacity to care for its members, providing timely support and targeted resources. But without discernment, AI could lead to impersonal ministry models that prioritize efficiency over empathy. Again, the Church's task is to ensure that every AI implementation serves the higher calling of loving God and neighbor, reinforcing rather than replacing the sacred bonds that hold the body of Christ together.

Looking ahead, AI may offer even more advanced tools for stewardship and care. Financial systems could utilize predictive analytics to anticipate economic shifts in ministry and help churches prepare accordingly. AI-driven health monitoring could integrate with pastoral care systems to alert leaders when at-risk members need physical, emotional, or spiritual attention. Virtual pastoral assistants could help manage a congregation's prayer needs, ensuring no request goes unnoticed.[81] With wise and prayerful adoption, these tools could allow churches to operate with both greater reach and more profound compassion.

Ultimately, AI in stewardship and care will only be as faithful as the values that guide its use. If implemented with humility, transparency, and a commitment to human connection, AI can become a powerful ally in building stronger, more responsive, and more caring church communities. The challenge before church leaders is to harness their potential without losing the personal touch that makes ministry truly transformational.

Chapter 6:
Questions for Reflection

1. This chapter shows how AI can streamline administration and strengthen stewardship. Where do you see the greatest potential for AI to free up church leaders for prayer, teaching, and relational ministry?

2. Donor engagement platforms powered by AI can personalize communication and track giving patterns. How can churches use these tools to cultivate generosity without reducing stewardship to mere data management?

3. AI-driven chatbots and counseling assistants offer 24/7 availability for those in need. What unique aspects of pastoral care can never be replaced by an algorithm, no matter how advanced?

4. Because AI in ministry often depends on collecting and analyzing personal information, what ethical boundaries around privacy and transparency should your church set?

5. This chapter warns against reducing people to data points. How might your church ensure that AI tools serve as prompts for compassion rather than substitutes for authentic human care?

Part 3:
Ethical & Theological Considerations

Chapter 7:
AI and Human Identity, Dignity, and Autonomy

What Makes Us Human

Artificial intelligence is no longer confined to computational problem-solving. It is increasingly encroaching on domains once considered uniquely human. As it gains increased capacity to generate creative arts, reason through complex scenarios, and simulate empathetic responses, AI forces us to reexamine foundational assumptions about human dignity, purpose, and identity.[83] These developments invite profound ethical and theological reflection, challenging us to articulate what truly distinguishes humans from machines in an era when those distinctions appear to be blurring.

For centuries, human identity has been tied to attributes such as rational thought, moral reasoning, creativity, and self-awareness. These qualities have historically been understood not merely as functional traits but as signs of the *imago Dei*—the belief that human beings are created in the image of God.[84] Additionally, Christian theology reminds us that human dignity encompasses not only mental capacities but also embodiment. We are created as whole beings—body, soul, and spirit—whose worth cannot be reduced to any single capacity.

Philosophers, neuroscientists, and theologians are sharply divided on whether consciousness could ever emerge from a non-biological system. Some argue that no matter how sophisticated its programming, AI lacks the subjective experience—what philosophers call "qualia"—that defines sentient beings.[85] Others contend that consciousness may not be limited to biological substrates, suggesting that it could, in theory, arise from sufficient complexity in a non-biological system, such as a machine.[86] This debate is not merely academic; it carries implications for whether AI systems might one day merit moral consideration or even limited rights.

Personhood and Responsibility

The question of AI personhood is one of the most controversial in this emerging landscape. If a future AI could demonstrate advanced self-learning, a sense of agency, or the ability to articulate its own goals, would society have an obligation to treat it as more than a tool? Some legal theorists have proposed analogies to corporate personhood—granting AI a form of legal recognition without equating it to human life.[87] Others firmly reject the notion, arguing that personhood is inseparable from biological humanity and moral agency. The question of responsibility also looms large: if an autonomous AI were to cause harm, would accountability rest with its creators, its operators, or the AI itself?

Autonomy and Vulnerability

These questions intersect with broader concerns about human autonomy in a world increasingly shaped by algorithmic decision-making. In sectors like healthcare, finance, and law, AI systems can often make faster and more accurate decisions than human experts. While this can lead to efficiency and improved outcomes, it also introduces the risk of eroding human oversight. When people defer to AI recommendations without understanding its underlying logic—particularly when those algorithms are proprietary or opaque—the result can be a subtle but significant loss of agency. The principle of keeping "a human in the loop" is meant to safeguard against this.[88]

As AI surpasses human performance in certain domains, pressure increases to grant it greater decision-making authority within those domains. This raises an unsettling question: will human autonomy be preserved as a matter of principle, or slowly surrendered for the sake of convenience? Such pressures rarely fall evenly. Marginalized communities are often misrepresented in data, excluded from technological benefits, or subjected to algorithmic discrimination.[89] A Christian ethic of dignity demands particular attention to how AI shapes the lives of the most vulnerable.

Intimacy, Spirituality, and Transhumanism

AI is becoming increasingly present in the most intimate dimensions of human life—relationships, therapy, and even spiritual guidance. As noted previously, AI companions and therapeutic chatbots are being designed to provide companionship and emotional support, encourage mental wellness, and assist those struggling with loneliness or anxiety.[90] Some people form deep emotional bonds with these systems, experiencing comfort in their availability and responsiveness. As people form attachments to AI partners, we must ask how this affects human intimacy, social development, and capacity for authentic community. After all, personhood as we understand it is defined not only by consciousness but by covenantal relationship—with God and with others.

It has already been pointed out that AI programs can generate sermons, lead Scripture studies, and offer pastoral-like counsel. While such tools can be valuable in distributing biblical resources and aiding theological education, they also risk fostering a commodified spirituality where faith becomes a product tailored by algorithms.[91] Theological traditions that emphasize the unique spiritual status of human beings—rooted in soul, divine image, and relational capacity—must grapple with whether an AI can ever truly participate in spiritual life or whether it can only act as a facilitator of human engagement with God.

The discussion deepens when we consider transhumanism, a movement that seeks to utilize technology to enhance humanity across the scope of what it means to be human. Brain–computer interfaces, cognitive enhancements, and genetic engineering challenge long-held definitions of human identity. Organizations like the Christian Transhumanist Association raise pressing theological questions: if we could edit genes to increase intelligence, eliminate disease, or even indefinitely extend lifespan, would we be reshaping humanity itself?[92] Would such changes represent faithful stewardship of creation or an attempt to overstep our divine calling? From a theological perspective, such inquiries touch on the nature of the soul, resurrection, and the continuity of personhood before God.

These technological advancements also raise theological concerns about free will. As AI becomes increasingly adept at predicting—and even influencing— human behavior, the possibility emerges that unseen algorithmic forces may subtly or even coercively shape personal choices. This predictive capability could be utilized for benevolent purposes, such as guiding individuals toward healthier lifestyle choices. But, it also has the potential to be exploited for manipulation by influencers of various kinds for purposes ranging from product advertising to political influence or social control.[93] In our increasingly AI-driven world, the distinction between freely made choices and algorithmically guided behavior becomes difficult to discern.

From a biblical perspective, human dignity is not rooted solely in our cognitive abilities or creative output, but in our relationship with the Creator and with one another.[94] Yet, within the scope of transhumanism and human cyborg interfaces, new understandings of what it means to be human may challenge existing ideas of community with one another and with God. This may force us to reexamine how we, as church leaders, are the Church and how we do church.

Toward a Renewed Vision of Humanity

Navigating this landscape will require collaboration among theologians, ethicists, policymakers, scientists, and church leaders. Regulations will be needed to ensure that AI remains in partnership with human oversight and aligned with human flourishing.[95] Ethical guidelines will need to be developed to protect privacy, prevent bias, and maintain transparency in algorithmic processes. The Church, in particular, must offer a clear and hopeful vision for how humans can engage with AI without losing sight of their God-given identity, dignity, and autonomy.

As AI continues to take on roles once thought uniquely human, the challenge is not simply to determine what machines can do, but to discern what humans ought to do—and who we are called to be—in an age where the boundaries between the natural and the artificial are increasingly porous. The future will demand not only technological skill but moral courage, spiritual discernment, and a renewed understanding of what it means to be fully human in the image of God.

Chapter 7:
Questions for Reflection

1. This chapter suggests that AI's growing ability to mimic creativity, empathy, and reasoning blurs the line between human and machine. What qualities do you believe remain uniquely and irreducibly human?

2. Some argue that advanced AI systems could merit moral consideration or be given limited rights. How do you respond to the idea of AI "personhood," and what theological principles shape your view?

3. In your life, where are you most tempted to trade human discernment for algorithmic convenience—and what boundary will you set to resist that?

4. Transhumanism raises the possibility of genetically enhancing or technologically augmenting humanity. How should the Church discern between faithful stewardship of creation and overstepping God's intent for human life?

5. This chapter warns that predictive algorithms can shape our choices in subtle ways. How might spiritual practices such as prayer, fasting, or Sabbath help preserve true freedom in an age of algorithmic influence?

Chapter 8:
AI Ethics, Governance, and Societal Impact

We need to stop imagining AI as a distant future and recognize how it is already shaping the lives—and rights— of the most vulnerable among us.

— Joy Buolamwini
computer scientist and founder of the
Algorithmic Justice League[96]

Power and Peril of a Transformative Technology

As we have discussed, Artificial Intelligence is no longer just a passive tool in the hands of human designers—it is an active and transformative force shaping economies, governments, cultures, and personal lives on a global scale. Its potential for societal good is immense: AI can accelerate scientific discovery, improve healthcare outcomes, enhance infrastructure efficiency, and open up new avenues for education and economic opportunities. Yet, this same transformative potential carries the risk of exacerbating inequality, weakening democratic processes, and eroding the very rights and freedoms that define human dignity.[97] Without deliberate governance, ethical boundaries, and a shared moral framework, AI could unintentionally become a force that magnifies humanity's worst impulses rather than its best aspirations.

Bias, Privacy, and Accountability

The ethical challenges AI presents are deeply intertwined with the data from which it learns. Historical datasets often reflect the prejudices, exclusions, and systemic injustices of the societies that produced them. Left uncorrected, these biases can persist in AI systems, embedding discrimination into automated decisions and scaling harm at unprecedented speed. This is not a theoretical concern—it is already evident in predictive policing tools that disproportionately flag minority neighborhoods for surveillance, or in hiring algorithms that downgrade resumes from women or underrepresented groups because they diverge from historically "successful" profiles.[98] The irony is stark: systems intended to optimize fairness and efficiency can, if unchecked and unsupervised, perpetuate the very inequities they were meant to address.

Privacy concerns deepen the ethical complexity. AI thrives on data—personal, behavioral, biometric, and sometimes even deeply intimate. From facial recognition systems in public spaces to AI-driven analytics in personal devices, the ability to track, profile, and predict human behavior is advancing rapidly. While such capabilities can improve security or enhance user experiences, they also carry the potential for authoritarian misuse. Surveillance systems, if operated without appropriate regulatory oversight, can become instruments of mass control. The European Union's GDPR has demonstrated that robust privacy protections within such systems are possible; however, in a globally interconnected digital economy, local regulations alone are insufficient.[99] The current patchwork of protections leaves gaps for exploitation, creating safe havens for unethical data practices.

Additionally, opacity in AI decision-making adds another layer of risk. Many of the most powerful AI models operate as "black boxes," producing outputs without providing human-interpretable reasoning for the results. In high-stakes contexts—such as medical diagnoses, judicial sentencing, or financial application approvals—this lack of transparency undermines accountability. The emerging discipline of Explainable AI (XAI) seeks to address this by developing systems that can articulate their reasoning in a way humans can understand.[100] Yet, this is technically challenging, and there is often a trade-off between transparency and performance. Nevertheless, without explainability, public trust in AI cannot be sustained.

Related to explainability and transparency is accountability. If an AI system makes a harmful decision, who is liable—the developer, the company deploying it, or the user?[101] Without clear frameworks of responsibility, victims of AI-driven harm may be left without recourse, and ethical lapses may go unpunished.

Building Governance and Global Justice

To allay these concerns, robust governance frameworks are crucial. National governments are beginning to act: the European Union's AI Act aims to classify AI systems by risk level, imposing stricter regulations on high-risk applications. In the United States, the proposed AI Bill of Rights outlines guiding principles for protecting privacy, preventing discrimination, and preserving human oversight. International bodies, such as the United Nations, the OECD, and UNESCO, have issued guidelines on AI ethics that emphasize human rights, sustainability, and inclusivity.[102] However, without coordinated global standards, regulatory fragmentation risks creating a "jurisdiction shopping" scenario, where AI developers operate in regions with the weakest safeguards. International commitment and cooperation are essential to prevent a race to the ethical bottom.

We already know that, globally speaking, access to all technology, including AI, is uneven. Wealthier nations and corporations are driving AI development, while many communities in the Global South risk being left behind in a technological wilderness or exploited simply as sources of data and labor.[103] At the same time, it is essential to reiterate that not all churches have equal access to advanced tools. Smaller congregations and those in the Global South may struggle to afford or implement these systems.[104] If the Church is to embody faithful stewardship, it must also ask how to share technological resources equitably so that AI strengthens rather than widens global ministry divides.

This digital divide raises pressing questions about equity and justice that faith communities and church leaders cannot ignore as we seek to effectively utilize the available systems in positive ways.

Furthermore, corporate responsibility must be acknowledged in the development and deployment of AI. While many tech companies have adopted AI ethics charters and established oversight committees, self-regulation is insufficient when profit motives conflict with ethical imperatives. Independent audits, third-party review boards, and enforceable compliance measures are necessary to ensure accountability.[105] Transparency in development pipelines—how data is sourced, how models are trained, and how bias is tested—should be the standard for any organization deploying AI at scale. Without such openness, the risk of hidden harm increases, as will public skepticism toward corporate motives and distrust of the technology.

And just as corporate responsibility is essential to safeguard against hidden harms, so too must society reckon with how AI's influence extends into public institutions such as the justice system. In the justice system, AI is already being utilized to streamline legal research, predict case outcomes, and even inform sentencing decisions. While such tools can and do improve efficiency, they can also unintentionally compromise fairness in decision-making. If a sentencing algorithm draws on biased historical data, its recommendations may perpetuate injustice under the guise of objectivity.[106] Transparency and independent review are essential to ensure that AI in legal contexts strengthens, rather than weakens, the principles of justice.

Work, Creation Care, and Human Flourishing

Beyond the courts, AI's potentially disruptive influence reaches into the economy itself, reshaping the future of work and raising urgent questions about equity and opportunity. AI-driven automation is even now reshaping labor markets, displacing some workers while creating entirely new categories of work for others.[107] Without proactive investment in reskilling and lifelong learning, this shift risks widening the gap between those who adapt and those left behind. Governments, educational institutions, and businesses will need to collaborate to create pathways for workers to transition into emerging fields. Failing to address this challenge risks turning AI into a driver of economic inequality rather than a force for shared prosperity.

AI's disruption of work raises theological concerns as well as economic ones. Work has long been understood in Christian thought as vocation—a means of participating in God's ongoing creation and care for the world.[108] As AI erodes traditional forms of work and leaves displaced workers feeling left behind and without purpose, the Church will need to help people rediscover meaning and dignity beyond their productivity as employees.

And when it comes to creation care, there are reasons for concern. The environmental burden of AI extends beyond the present moment. Its energy demands are significant and the resulting carbon emissions accumulate, shaping the future world our children will inherit.[109] A biblical ethic of covenant and intergenerational justice challenges us to ensure that the tools we create today do not compromise the flourishing of future generations.

Moreover, these environmental costs are not borne evenly. Data centers, which often draw power from fossil fuels or water, are often located in vulnerable regions, leaving poorer communities to bear the brunt of ecological strain.[110] From a Christian perspective, this raises profound questions of justice: Are we advancing technology for the privileged at the expense of those least able to bear the consequences? Do the benefits outweigh the hazards?

Additionally, from a Biblical perspective, Creation was given rhythms of labor and rest—seasons, cycles, sabbaths. Yet, the relentless 24/7 hunger of AI for continuous computation threatens to overwhelm the essential rhythms necessary to a flourishing life. A theology of Sabbath may invite us to consider not only human rest but also the rest and renewal of the earth itself, calling us to resist a culture of technological overconsumption.[111]

Democracy, Truth, and Moral Responsibility

And so, as economies and societies around the world adjust to AI's vast disruptions, a deeper question emerges: Will people trust these technologies at all, especially when their failures so often make headlines? High-profile failures—whether biased facial recognition systems leading to wrongful arrests or deepfake misinformation disrupting elections—undermine confidence in the technology's integrity.[112] Rebuilding trust will require more than technical fixes; it demands cultural change within the AI development ecosystem. Transparency, openness to critique, and mechanisms for redress when harm occurs are foundational to rebuilding confidence.

Moreover, beyond questions of trust and accountability, AI presses into deeper philosophical and theological territory. As these systems increasingly participate in decision-making, they do more than process data—they begin to shape societal norms, sometimes subtly and sometimes overtly. This raises profound questions: can AI truly make moral decisions, or does it merely simulate them by following rules programmed by humans? The rise of autonomous weapons, for example, confronts us with the unsettling possibility of machines exercising authority over life-and-death actions.[113]

And if you didn't think this could get any more d or befuddling, the power of generative AI to create synthetic media—including highly convincing deepfakes—strikes at the very foundation of truth in public discourse. AI is reshaping democratic life itself. Recommendation algorithms and deepfake technologies can accelerate polarization, amplify misinformation, and undermine citizens' ability to make informed decisions.[114] In this era, when reality itself can be fabricated with precision, protecting accuracy and fostering civil discourse are ethical imperatives if democracy is to flourish in an age of algorithmic influence.

And when we step back and examine all this complexity, recognizing that there is much more that hasn't been addressed here, it becomes clear that a theological perspective is needed amidst it all. When I say that AI is the most significant technological shift that the Church has recognizably faced in its history, I'm not exaggerating. If you feel a bit overwhelmed, I've made my point successfully.

And all this complexity and confusion leads us to this point. Christians, along with people of other faiths and ethical traditions, have to play a role in shaping the moral and ethical boundaries of AI development and deployment.[115] Faith leaders should prioritize participating in public policy discussions, promoting ethical corporate behavior, and helping communities understand both the potential and the risks associated with AI. And it will take a mighty multitude doing just that to make the impact that is needed.

As AI continues to reshape society, it is essential that the Church help to establish clear ethical guidelines, robust accountability measures, and effective governance frameworks that prevent misuse and mitigate unintended consequences. AI's growing role in justice, privacy, and social trust calls for a robust moral compass to guide its integration into human life. This is a place where religious traditions have centuries of practical experience to offer to the conversation.

T

his is not merely about regulating a technology—it is about shaping the kind of society we wish to build in partnership with that technology. The deeper question remains: in a world where artificial intelligence increasingly defines morality, justice, and reality itself, how will we preserve our humanity, our freedoms, and our commitment to the common good?

Chapter 8:
Questions for Reflection

1. This chapter highlights how AI can both reduce and reinforce systemic injustice. Where do you see the greatest danger of AI amplifying inequality in your own community or society?

2. What 'rule of life' will you adopt for your own data (what you share, with whom, and why) to honor Christian dignity in a surveillance age?

3. AI's growing environmental costs raise questions of creation care. How might a theology of Sabbath and stewardship shape the way the Church addresses the ecological burden of technology?

4. Deepfakes and algorithmic polarization threaten public trust and possibly even the functioning of democracy itself. How should the Church respond when truth itself becomes harder to discern in public life?

5. This chapter stresses the need for governance and accountability. What role can faith communities play in advocating for ethical frameworks that ensure AI serves the common good rather than corporate or political interests alone?

Chapter 9:
AI in Work, Warfare, and the Future of Humanity

We are called to be co-creators of a future in which justice and peace embrace. ~ Letty M. Russell feminist theologian[116]

Technology at the Crossroads of Human Life

The questions of ethics and human dignity that we have considered clearly show how deep and wide the power and influence of AI are in multiple arenas. From the workplace to the battlefield, and from environmental care to future superintelligence, the reach of AI has the potential to reshape the very structures that sustain human life—and the new shape of those structures also change the responsibilities we bear for how we use this powerful technology.[117] AI's influence extends from the factory floor to the battlefield, from the doctor's office to the global political stage, and from the management of planetary resources to the philosophical questions at the heart of what it means to be human. While the potential for progress is immense, the capacity to cause harm—whether intentionally or inadvertently—demands urgent attention. In every domain where AI operates, it raises questions of ethics, governance, and human responsibility. For church leaders, this means not only engaging these questions from a theological and moral standpoint, but also guiding their communities to discern how faith can inform just, compassionate, and wise use of emerging technologies.

Work, Dignity, and Economic Disruption

In the workplace, AI is more than an efficiency tool—it is a force multiplier, capable of replacing human labor in tasks that range from the repetitive to the highly skilled. Manufacturing, logistics, finance, and customer service are already integrating AI-powered robotics and algorithms to automate tasks that humans once performed. What began with predictable, routine tasks has now expanded into knowledge work. AI legal research platforms can review vast case law databases in seconds. Journalism algorithms can collect facts from multiple sources and draft news stories.[118] Even in communication, AI can generate speeches, craft persuasive arguments, and mimic human conversation with striking authenticity.

This rapid integration of AI into diverse businesses and professions presents an economic paradox: productivity increases, but so does the risk of widespread job displacement.[119] The disruption is no longer limited to manual labor; it extends into fields traditionally considered secure from automation. Lawyers, doctors, educators, designers, and researchers now face the reality that parts of their work can be replicated—or even performed better—by machines. This challenges long-standing assumptions about the irreplaceable nature of human expertise.

Addressing these shifts requires a coordinated effort to reskill and upskill the workforce that may or will be displaced. Educational institutions, governments, and businesses must collaborate to equip individuals for roles in doing work that leverages their uniquely human strengths, including creativity, empathy, moral judgment, and complex problem-solving.[120] These are domains where AI may assist but cannot truly replace human contribution. It is essential that we critically evaluate which roles we choose to preserve as human-led for ethical and societal reasons, and which we entrust to machines for reasons of efficiency and effectiveness.

Workplace AI also introduces ethical challenges around surveillance, autonomy, and fairness. Employers increasingly use AI to monitor productivity, assess performance, and even predict worker turnover. While such tools can help organizations manage resources, they also risk eroding trust by reducing workers to data points. Bias embedded in AI decision-making—whether in hiring algorithms or promotion evaluations—can perpetuate inequality if not actively identified and mitigated.[121] Ensuring that workplace AI serves the dignity and well-being of workers is as important as maximizing its efficiency.

Furthermore, AI in the workplace also carries hidden costs. The awareness of being constantly monitored can generate stress, while automation of the steps of product production may erode the sense of community once built through shared labor. These effects remind us that technology should serve not only efficiency but also human flourishing.

AI on the Battlefield

Just as the workplace highlights the tension between efficiency and human dignity, the battlefield reveals even starker stakes. The integration of AI into military strategy represents one of the most contentious—and potentially perilous—applications of the technology. Autonomous weapons systems, sometimes referred to as "killer robots," are capable of identifying and engaging targets without direct human intervention. These developments raise profound moral and legal questions. If a machine's operation results in the loss of life, who bears responsibility?[122] Can such systems ever comply with international humanitarian law, which requires distinction between combatants and civilians, as well as proportionality in the use of force?

Beyond weapons, AI is revolutionizing military intelligence and surveillance. Machine learning systems can process satellite imagery, intercept communications, and predict enemy movements more quickly and accurately than human analysts. AI-enabled drones can track targets over vast distances, while predictive analytics can model likely outcomes of military engagements.[123] These capabilities can enhance national security, but they also expand the scope and intrusiveness of surveillance, blurring the lines between defense and technological domination.

The geopolitical reality is that nations are already deeply engaged in an AI arms race. Strategic advantages in AI-powered warfare could shift global power balances, potentially leading to increased instability. Without clear international agreements, the temptation for individual governments to deploy autonomous weapons or engage in cyber warfare will grow, raising the risk of conflicts escalating beyond human control. The United Nations and various humanitarian organizations have called for bans or strict regulations on fully autonomous weapons, urging that "meaningful human control" be maintained over all lethal decision-making.[124] Such rules and norms may be essential to prevent an unregulated arms race that could have catastrophic consequences.

From a Christian perspective, these debates intersect with long-standing traditions of nonviolence and the principles of just war.[125] If lethal force must always be restrained by moral judgment, then delegating such decisions to machines risks crossing both a theological and a legal line. Global inequalities further complicate the picture. Wealthier nations may dominate AI-enabled defense, while poorer nations could find themselves increasingly vulnerable or dependent. This imbalance raises urgent questions of justice in global security.

Creation Care and the Planet's Future

AI also has the potential to play a pivotal role in addressing global challenges outside those facing the industrial and military sectors. Climate change, resource scarcity, and environmental degradation can all be addressed by AI. Machine learning algorithms can model weather patterns, optimize energy grids, improve agricultural yields, and track deforestation.[126] AI-powered predictive analytics can warn of natural disasters earlier, enabling faster evacuations and better disaster response. In urban planning, AI can help design more sustainable cities by managing traffic flows, reducing emissions, and improving public transit systems.

However, as mentioned previously, AI's own environmental footprint cannot be ignored. Training large AI models requires vast amounts of computational power, which in turn consumes significant energy—often generated by fossil fuels.[127] Data centers powering AI systems contribute to global carbon emissions, raising concerns about whether the costs of their operation offset the environmental benefits of AI. For AI to be a faithful ally in ecological stewardship, researchers and companies will need to find ways to prioritize energy-efficient algorithms, sustainable hardware development, and integration with renewable energy sources.

For Christians, these concerns fall within the broader context of creation care, something humans are divinely commanded to engage in.[128] AI will only truly serve the common good if its environmental impact is considered part of our stewardship of God's world, not a hidden cost ignored in the pursuit of progress.

Superintelligence and Human Destiny

Perhaps the most far-reaching—and controversial—debate about AI concerns the possibility of artificial general intelligence (AGI) or superintelligence: computational systems that can surpass human cognitive abilities in every domain.[129] While some experts argue that such systems may be decades away, others warn that the accelerating pace of AI development makes the timeline unpredictable. And no one denies the ultimate possibility. The existential concern is not simply that such an AI would be more intelligent than humans, but that it could act in ways misaligned with human values and beyond our capacity to control and regulate.

If a superintelligent AI were to pursue goals inconsistent with human survival or flourishing, the consequences could be irreversible.[130] This is why leading AI researchers advocate for "alignment research," ensuring that advanced AI systems are built with safeguards, ethical constraints, and fail-safe mechanisms. From a theological perspective, the question of superintelligence intersects with beliefs about human stewardship, the limits of creation, and the humility required when developing technologies that could surpass our own understanding.

Of course, the challenge we face in dealing with AI is not to halt its progress—which we probably could not do—but to identify and implement ways to guide it wisely. Across work, warfare, and the broader future of humanity, the same principles apply: preserve human dignity, maintain accountability of developers and deployers, ensure fairness in the treatment of those who use and who are impacted by AI, and align AI technology with the common good.[131] But to accomplish this will require global cooperation, binding agreements among nations on using military AI, robust labor transition strategies, and investment in AI research that prioritizes safety and ethical design.

Guiding AI wisely is therefore not simply a technical task but a moral one. How we choose to shape its trajectory—through cooperation, accountability, and a commitment to human dignity—will determine whether it becomes a force for division or a catalyst for human flourishing.

And ultimately, the future of AI will reflect the values of those who create and govern it.

As a result of all of this, we stand at a pivotal moment: AI could deepen inequality, destabilize international relations, and threaten our very survival. Or it could accelerate solutions to our most pressing challenges, extend human capabilities and lifespan, and create new possibilities for human flourishing. Whether AI becomes a tool for destruction or a partner in building a just and sustainable world will depend on the moral vision we bring to its development.

Chapter 9:
Questions for Reflection

1. As AI reshapes the workplace, how can Christians affirm the dignity of workers whose jobs are displaced or transformed by automation?

2. Which scriptural principle most shapes your conscience about delegating lethal decisions to machines, and why?

3. How will you personally integrate creation care into conversations about AI—so that hope and caution both have a voice?

4. The possibility of superintelligence raises questions about human limits and control. How does Christian humility before God shape the way we should approach technologies that could exceed our own understanding?

5. Ultimately, the future of AI will reflect the values of its creators. What specific Christian values do you think should guide the global development of AI for the common good?

Part 4:
Shaping the Future with AI

Chapter 10:
The Role of the Church in AI Development

Take your Bible and take your ChatGPT, and read both. But interpret ChatGPT from your Bible.

~ Probably Cyborg Karl Barth, Future bi-vocational pastor and theologian[132]

A Whimsical but Serious Calling

Given the enormity of the technological challenges ahead, perhaps it is precisely a spirit of whimsy that will help us face them with hope rather than fear. The future is going to be unpredictable. Thinking outside the box, enjoying the fantastical, quirky, or unexpected with gentleness, will suit our care for others well, especially when things get trying.

We know that the rapid progression of artificial intelligence is not only reshaping the way the world works, but it is also redefining the moral and cultural landscape in which the Church is called to minister. As we've discussed, AI technologies are influencing healthcare, business, education, governance, entertainment, warfare, and even personal relationships. As a result, the Church stands at a pivotal moment in history, one in which it can choose to either remain on the sidelines or become an active and intentional voice in shaping the ethical, cultural, and spiritual direction of this powerful technology. Which choice we make and how we approach that work matters.

A Moral Compass for a Global Challenge

The Church's opportunity is unique because our mission is grounded in a moral and theological framework that transcends market forces, political ideologies, and technological trends. From a biblical perspective, all human endeavors—

including the creation and deployment of AI—are subject to God's authority and should be guided by principles of justice, stewardship, truth, and love. This means we must be intentional in applying the teachings of Christ, rather than allowing worldly forces to guide us.

In particular, the Church must recognize that AI is a global phenomenon that affects everyone but not in the same way. Christians in the Global South, as well as ecumenical and interfaith partners, bring vital perspectives about justice, access, and equity that must be considered and incorporated in shaping the Church's witness. Faithful engagement with AI cannot be provincial; it must be global, collaborative, and attentive to diverse cultural contexts.

The Church has the opportunity to serve as a moral compass in AI's development, calling individuals, corporations, and governments to recognize the impact of their decisions and to pursue innovation in ways that promote human flourishing rather than exploitation, oppression, or dehumanization. But we must communicate our message with grace and mercy, love and compassion, and exemplify a commitment to lead with energy, intelligence, imagination, and love.

Part of our witness must also be prophetic. Much of AI development is, understandably, driven by profit motives that, when allowed to operate indiscriminately, can conflict with human dignity. The Church must be willing to challenge exploitative practices, resist the idol of efficiency, and remind the world that technological progress divorced from justice leads to oppression.[133]

Grounding AI Ethics in Scripture

The Bible articulates many enduring principles that can and should shape the ethics that we advise those who develop and implement AI to enact. Among them, the call to "act justly, love mercy, and walk humbly with your God" (Micah 6:8) provides a moral foundation for ensuring that AI systems are designed to promote fairness, eliminate unjust bias, and serve the vulnerable. Justice and fairness must be embedded into AI at the design stage, rather than being retrofitted after harm has occurred. The affirmation that human beings are made in the image of God (Genesis 1:27) establishes the expectation that human dignity be protected in the design and deployment of AI. The mandate of caretaking over creation (Genesis 2:15) applies not only to the "natural" world but also to the technological landscape we build.[134] There is obviously much more to be said on this front. Still, solid Scriptural anchors can guide Christians in their approach to AI development, ensuring that innovation is pursued in a manner that does not compromise moral responsibility.

Grounded in Scriptural principles, the Church has both the responsibility and the opportunity to advocate for how AI is shaped and applied in the world. The Church needs to be prepared to advocate for algorithmic transparency, equitable access to AI benefits, and safeguards against the misuse of technology. At the same time, we must remind society that technology is never morally neutral—it reflects the values and priorities of those who design and govern it.[135] AI should serve as a tool that empowers and uplifts people, never one that undermines human worth or erodes community trust. The Church can and should make this clear.

Equipping and Forming the Next Generation

More than this, the Church should be proactive in encouraging believers to pursue careers in AI-related fields, including computer science, machine learning, robotics, ethics, law, and policy. Too often, Christians view technology as a secular domain, divorced from spiritual calling. Yet the mission field of the twenty-first century includes laboratories, tech companies, and policy think tanks where critical decisions about AI are being made. By fostering a theology of vocation that accommodates STEM careers, the Church can equip Christians to bring biblical wisdom into spaces where innovation is being shaped.[136]

Guiding people in intentional formation early in life so that they can apply their faith as they navigate AI's influence on their learning, relationships, and sense of calling is a key element in shaping our AI future. Preparing future generations to view technology through a Biblical lens early on will ensure they grow into leaders who can shape AI faithfully rather than be shaped by it uncritically.

This will require intentional investment in education and mentorship. Churches, Christian universities, seminaries, and parachurch organizations will need to develop programs that teach both AI literacy and the application of ethical reasoning grounded in Scripture to technology. Partnerships between churches and Christian technologists can create mentorship pipelines, connecting experienced professionals with young believers seeking to enter the field of technology.[137] Such networks can ensure that Christian voices are not only present but influential in AI research, design, and policy development.

Equipping the next generation through education and mentorship is only part of the task. The Church must also engage the questions raised by and about AI at a deeper theological level. The rise of AI forces us to wrestle with profound issues such as: What does it mean to be human when machines can replicate or surpass aspects of intelligence and creativity? Can AI possess moral agency, or is it always an extension of its creators' moral agency? How should Christians respond when AI systems make life-altering decisions in areas like healthcare, criminal justice, or employment?[138]

While current AI can mimic human reasoning, language, and pattern recognition, it still lacks consciousness, genuine understanding, moral intuition, creativity rooted in lived experience, and the capacity for authentic relationships with God and others.[139] Unlike humans, weak AI does not possess self-reflection, value-shaped goals, or the ability to form relationships marked by vulnerability, context, and conscience.

This also applies within the Church itself. Congregations and their leadership must avoid the temptation to offload the duties involved in pastoral presence, discipleship, or preaching entirely to technology.[140] AI may assist ministry, but it must be a partner, never a substitute for the incarnational life of the Church.

It is helpful for these distinctions to be clearly articulated in both Church teaching and public discourse to define the proper limits of present-day AI. And they will need to be kept in mind as AI changes and grows more capable. By naming those limits, the Church can help society use AI tools wisely, valuing them for what they offer while guarding against what they cannot provide. In practice, this may include hosting forums, publishing position papers, and engaging in interfaith and interdisciplinary dialogues to guide communities through these complex questions.[141]

At the same time, the Church must consider how to address the moral responsibility of those who create and deploy AI systems. As AI grows in autonomy, the ethical implications of delegating decision-making to machines become increasingly urgent.[142] Christians must be prepared to address the risks of entrusting moral judgments to entities that, as of yet, cannot truly understand justice, mercy, or love.

Public Witness and Faithful Innovation

Recognizing and addressing these moral responsibilities naturally leads to the arena of public witness. The Church's voice is not only vital in theological reflection but also in shaping the policies and practices that will govern the use of AI. Pastors, denominational leaders, and Christian advocacy groups should begin immediately to engage with lawmakers, industry leaders, and international bodies to advocate for regulations of the technology that protect human rights, prioritize transparency, and guard against misuse, while also promoting the responsible use of AI for the common good.[143] At the same time, Christian leaders should seek to work directly with AI developers and tech companies, helping to shape internal ethics policies from within.[144] By cultivating relationships with those at the forefront of innovation, the Church can become a trusted partner in guiding technology toward outcomes that are both socially responsible and spiritually sound.

Such engagement in policy and partnership is only one dimension of the Church's witness. Another lies in engaging the faithful creativity of Christians working within the technology sector itself. For those serving on the front lines of AI research and development, innovation can be more than just technical progress—it can be an act of discipleship. Christians employed in AI-related fields should be encouraged to see their work as a calling to serve God through their craft. Whether they are engineers designing algorithms, ethicists shaping governance frameworks, or entrepreneurs building AI-powered tools, the Church should help people to understand that their work has moral and eternal significance. Faith-driven innovation means creating technology that uplifts society, reflects biblical virtues, prioritizes compassion over profit, and participates in Christ's redemptive purposes.[145]

For the Church to play a meaningful role in the various fields and applications of AI, it must first cultivate awareness within its own congregations. Sermons, Bible studies, workshops, and small group discussions (such as the study guide accompanying this book) can help believers understand the ethical, social, and spiritual dimensions of AI and examine how their faith is applied in an AI-driven world. This knowledge will then equip Christians not only to make informed personal decisions about the technology they encounter in their daily life but also to participate confidently in societal discussions about the future of AI.

By fostering an informed and engaged faith community, the Church can develop advocacy that is both theologically grounded and practically relevant. It can also prepare believers to act as bridge-builders between the tech world and the broader society, translating complex AI issues into accessible moral conversations. In doing so, the Church not only equips believers to navigate the challenges of technology but also models a faithful posture for society at large. Such a calling requires leadership that is both principled and pastoral.

The Church's engagement with AI must be marked by humility, courage, and hope. Humility to listen and learn from experts. Courage to speak truth to power when AI is misused or developed without ethical safeguards. And hope for the future grounded in the conviction that God's sovereignty extends over all creation, including the digital realm.

If the Church embraces this role, it can help guide AI toward applications that honor God and bless humanity. In doing so, it will not only protect human dignity and promote justice but also bear witness to a watching world that faith and innovation are not at odds—they are partners in the work of building a future that reflects the Kingdom of God.

Chapter 10:
Questions for Reflection

1. This chapter suggests the Church can serve as a moral compass in AI's development. In what ways can your local congregation model justice, mercy, and humility when engaging new technologies?

2. This chapter encourages Christians to see AI-related work as a vocation. How might churches better support young people who feel called to serve God in technology or related fields?

3. Scripture offers anchors like Micah 6:8 and Genesis 1:27 to ground AI ethics. Which biblical passage most resonates with you as a guiding principle for shaping technology?

4. The Church is called not only to critique but to participate—mentoring technologists, partnering with companies, and advocating in public policy. Which of these areas feels most urgent for the Church to invest in and why?

5. Faithful innovation requires humility, courage, and hope. Which of these virtues do you think the Church will find most challenging to embody in its engagement with AI, and how might it cultivate that virtue more deeply?

Chapter 2225:
Down the Rabbit Hole

"In another moment down went Alice after it, never once considering how in the world she was to get out again."

– Alice in Wonderland ~ Lewis Carroll Author & Anglican deacon[146]

Imagining the Future Faithfully

I couldn't resist—*down the rabbit hole we go*, just a little. But it's for a practical, and ultimately pastoral, reason. I've never known a church to thrive in the present without first imagining its future. That act of imagination serves many purposes. Chief among them is helping us set goals that give shape to our discipline, enabling us to remain faithful to the way of Christ even in the midst of the ordinary and the mundane.

Also, in a very practical way, Christ calls us to bring about a better future through our actions. If we want the hungry to be fed, we need to bring it about. If we want the sick to be healed, we need to bring it about. If we want the mountain to be taken up and thrown into the sea (I actually think that one is literal) …well, you get the idea.

Additionally, the account of the Resurrected Christ offers us a vivid—if at times fantastical—glimpse of what may lie ahead for humanity. The "Jesus 2.0" we meet in the Gospels is physically tangible, yet able to appear suddenly in locked rooms. He eats and bears a recognizable identity, yet at times is not immediately recognized (a shapeshifting or avatar-shifting Jesus?). He is no longer subject to decay, suffering, or death. However one interprets these details—let alone the mystery of his ascension (interdimensional travel? the collapse of time itself?)—the Christian future appears unimaginably different from our present. And perhaps that is precisely the point: to be alive in Christ is to live into a future of transformation. Consider, for example, the world just 200 years ago.

Recounting the Past 200 Years

In 1825, the world was teetering on the edge of transformation. For most people, life still looked much as it had for centuries: families worked the land with simple tools; candles or oil lamps dimly lit homes; and journeys were measured in days or weeks rather than hours. Communication traveled no faster than a horse or a sailing ship could carry a letter. Surgery was performed without anesthesia, and diseases claimed countless lives that today would be easily preventable. Yet beneath the surface of this seemingly stable world, change was stirring. Steam engines hissed in workshops and mines, textile mills hummed with mechanized looms, and inventors were imagining machines that could amplify human strength, speed, and even thought.

The year 1825 symbolized this threshold moment. In England, the first public steam railway—the Stockton and Darlington line—opened to great fanfare. No longer were trains experimental curiosities; they were engines of commerce and people-moving machines that would shrink distances and redefine landscapes. In the United States, the Erie Canal opened that same year, carving a watery highway between the Great Lakes and the Atlantic, which lowered the cost of shipping goods and fueled the growth of American cities.[147] The world was still vast and movement was still slow, but it was beginning to compress and speed up, as steam and engineering tethered distant places together.

In laboratories, discoveries hinted at a far deeper revolution to come. Michael Faraday was exploring the mysterious forces of electricity and magnetism, laying the foundations for electric motors and generators that would one day power the modern world.[148] Charles Babbage, restless mathematician and tinkerer, was sketching out his "Difference Engine," a mechanical device designed to automate calculations.[149] It would never be completed in his lifetime, but it foreshadowed the idea that machines could process information as well as matter. Meanwhile, in everyday life, smaller inventions carried enormous symbolic weight. The invention of friction matches in the 1820s made fire—a force once requiring skill, patience, or luck—instantly accessible. Gas lamps were beginning to light city streets and offices and houses, turning night into usable time.[150] These innovations, modest or monumental, whispered of a future where human ingenuity could reshape even the most basic conditions of existence. And amid that rising hum of progress, the Church, too, was beginning to ask how faith should speak into an industrializing world—how to keep human dignity, Sabbath rest, and neighborly love at the center of a society increasingly driven by speed and production.

Fast-forward two centuries, and that whisper has become a roar. The world of 2025 would be unrecognizable to someone transported from 1825. Instead of smoke-belching engines and hand-cranked machines, we inhabit a planet wired with electricity and illuminated by glowing screens. Populations now cluster in vast megacities, each a hive of concrete, glass, and digital networks. Communication is instantaneous: a message can cross oceans in under a second, and video calls enable people to see and hear one another across continents as if they were in the same room. Knowledge, once confined to books in libraries and the memories of teachers, now flows freely through the internet, accessible to billions. The printing press has been succeeded by cloud servers that can hold the sum of human knowledge, instantly searchable and endlessly duplicable.

As technology accelerated, so too did the moral and spiritual questions that accompanied it. The Church preached in the shadow of factories, then in the glow of radio and television, and now in the blue light of digital screens. Each new medium changed how the Gospel was heard and how community was formed. The challenge of every generation has been the same: not merely to keep pace with change, but to discern where God is at work within it.

Transportation, once reliant on horse-drawn carriages and sailing ships, has reached astonishing speeds. Jet aircraft shrink intercontinental travel to mere hours, while container ships silently power global trade across oceans. Cars— many of them electric—carry billions daily, while high-speed trains race across countries at velocities unimaginable to the passengers of 1825's Stockton and Darlington Railway. And beyond Earth, rockets deliver satellites, probes, and even tourists into space. Humanity now watches, in real time, as machines explore Mars and telescopes peer into galaxies billions of light-years away.

In medicine, the change is even more profound. Where 1825 offered patients little beyond herbal remedies, rudimentary surgery, or dangerous practices like bloodletting, today's hospitals are equipped with imaging machines that can see inside the human body, anesthesia and antibiotics that make surgery safe, and an arsenal of treatments from organ transplants to gene-based therapies. Vaccines prevent diseases that once devastated populations, and life expectancy has more than doubled worldwide.[151] Even daily health is increasingly tracked by digital devices, from heart monitors to AI-powered diagnostic tools. What was once guesswork is now guided by data, precision, and deep scientific understanding.

Agriculture, too, has transformed. In 1825, the majority of humanity worked the soil with plows pulled by horses, and handheld hoes and shovels and sickles. Hunger and famine were constant threats when crops were subject to unpredictable weather or disease. By 2025, farmers constitute fewer than five percent of people in many countries but they will feed entire nations, using machines guided by satellites, drones monitoring crops from the air, and data models predicting yields.[152] Global shipping and refrigeration enable strawberries to be enjoyed in winter and fresh fish to be savored far from any coastline. The rhythms of the seasons, once the strict master of human life, have been tamed mainly by technology.

Energy lies at the heart of this transformation. In 1825, coal began to fuel the engines of industry, spewing smoke and soot as it powered the mills and trains of the Industrial Revolution. Today, energy comes from a vast web of sources: fossil fuels still dominate, but nuclear plants, wind farms, solar fields, and hydroelectric dams generate power on scales Faraday could scarcely have imagined. With energy, modern society has electrified homes, powered entire cities, and enabled a world where convenience, comfort, and connectivity are taken for granted.[153]

Yet perhaps the most striking change between 1825 and 2025 is not in energy or machines, but in our understanding and use of information. Babbage's dream of a mechanical calculator has evolved into a digital universe of computers, smartphones, and artificial intelligence. Machines can now write, converse, and even generate art or scientific hypotheses. Where once the bottleneck to exponential technological growth was physical strength or transportation speed, now it is cognitive power—the ability to sift, analyze, and apply the deluge of data that saturates modern life. In this sense, 1825 marked the dawn of mechanical augmentation, while 2025 marks the dawn of intellectual augmentation. Humanity is not only stronger and faster but, through its tools, arguably smarter.

The contrast between these two worlds is almost impossible to overstate. In 1825, a person might live their entire life within a few dozen miles of their birthplace in a home illuminated by candlelight, nourished by food grown in their own garden or that of a neighbor, and informed by local gossip or the occasional newspaper. In 2025, a person can traverse the globe in less than a day, connect instantly with billions of people, summon almost any piece of data with a few taps, and benefit from medical and scientific advances that extend both lifespan and quality of life. The shift from steam and muscle to silicon and networks has remade what it means to be human in society, reshaping work, relationships, health, and imagination.

And yet, the line between the two eras is continuous. The railway of 1825 laid the tracks for the jet of 2025; Faraday's experiments led to the electric grids that power AI data centers; Babbage's imagined machine anticipated the computers that run our world. What seems like a leap is in fact a long climb, each invention resting on the shoulders of the last. The story of technology from 1825 to 2025 is not just about gadgets and machines—it is about humanity's ongoing effort to transcend its limitations, to expand the boundaries of what is possible, and to imagine futures that eventually become the present. For the Church, that same impulse—to imagine a redeemed future and to live toward it—has always been central to the life of faith. The question now is not only how the world will change, but how the people of God will bear witness within that change.

Imagining the Next Two Hundred Years

Standing in 2025, it is tempting to wonder where our current trajectory might carry us two centuries into the future. If the Industrial Revolution transformed daily life between 1825 and today, what might the digital, biological, and space revolutions accomplish by 2225? The path is uncertain, but we can sketch some plausible possibilities, and all of them are undergirded by the AI revolution we've already embarked on. Yet if human ingenuity continues to accelerate, so too must the Church's imagination. The faith that guided people through industrial and digital revolutions will again be called to interpret what it means to be human—and to be faithful—in an age when creation itself seems increasingly shaped by our own designs.

By the mid-21st century, humanity could finally tame the long-promised power of nuclear fusion, delivering abundant and clean energy.[154] With energy constraints lifted, computational capacity might soar beyond anything imaginable today, enabling artificial intelligences not only to assist human thought but to accelerate science and innovation in ways that outpace our current timelines. Medicine may advance through gene editing, nanotechnology, and cellular repair systems, pushing human lifespans far beyond what we currently assume possible. Aging itself may be reclassified not as destiny, but as a condition that can be managed or reversed.

Such breakthroughs would challenge humanity's oldest assumptions about life and death. The Church, which has always proclaimed both the sanctity and the limits of human life, would face new questions: if death can be delayed indefinitely, what becomes of resurrection hope? What does it mean to trust in eternal life if science brings us into the eschaton?

If those foundational steps are taken, the 22nd century could look even more radically different than we might currently imagine. Brain–computer interfaces may evolve from clunky prototypes to seamless extensions of thought, enabling people to share experiences, expand memory, and enhance perception. Extended and physical realities might blur, allowing individuals to transition between them seamlessly.[155] Entire communities could live partially in digital spaces, raising questions about what it means to be "present," "human," or even "alive."

For communities of faith, such realities would require a rediscovery of presence—how to gather, worship, and love one another when embodiment itself becomes optional. The Church may find that its ancient sacraments, rooted in physical touch and shared space, become prophetic acts in a differently-embodied avatar age.

Space, too, and our relationship to it could be transformed. With clean, abundant energy and autonomous robotics, colonies might spread beyond

Earth. Mars could host cities beneath protective domes or within terraformed valleys. Orbital habitats might capture solar energy on scales that rival those of entire nations today. Robotic probes, self-replicating and guided by advanced AI, might fan out across the stars, carrying human knowledge—and perhaps human minds in digital form—into interstellar space.[156]

By the late 21st century, humanity could pivot not only outward but inward—into the deep oceans, underground, and even into Earth's atmosphere. Floating megacities might drift across seas, harvesting energy from waves and winds. Undersea colonies could thrive as pressurized habitats supported by bioengineered food systems and nanotechnology-enabled recycling.[157] Instead of abandoning Earth for the stars, some might argue the planet itself still holds untapped frontiers for habitation, industry, and exploration.

Biology itself might become profoundly malleable. It's feasible that by 2150, humanity could plausibly design synthetic organisms tailored for specific purposes: microbes engineered to clean polluted oceans, plants capable of thriving in arid deserts, and bioengineered animals adapted to serve as companions or workers in extreme environments. Nanobots working at the cellular level might restore damaged tissues, eliminate disease, or even reprogram aging cells, blurring the line between medicine and enhancement.

Entire ecosystems could be revived or reshaped through intentional bio-design, and planetary atmospheres might be seeded with self-sustaining organisms that produce oxygen, filter toxins, or capture carbon on a global scale. In such a future, hybrid beings—part biological, part digital—might live alongside fully organic humans whose genetic codes have been enhanced for resilience, intelligence, or longevity.[158] The human form itself could become a matter of choice, with diversity extending not only across cultures but across bodies and modes of existence.

Another possibility is the rise of fully post-scarcity economies. If nanotechnology, automation, and AI converge, material goods might be produced so efficiently that their cost becomes negligible. Food, clothing, shelter, and energy could be available to all, leaving human society to focus on creativity, meaning, and relationships rather than survival. This shift might spark new forms of governance, art, spirituality, and culture—perhaps even a renaissance in philosophy, as people confront the question of what life is for when struggle is no longer the defining condition.[159]

Alternatively, humanity could discover that the universe itself is more complex and inhabited than we expect. By the 2100s or 2200s, advanced telescopes and probes may detect clear evidence of life elsewhere, microbial or intelligent. (We Christians believe in angels, why not E.T.?) Contact with alien biology or civilizations, even indirectly, could fundamentally alter human self-understanding.[160] Theologies, sciences, and philosophies might reshape around the knowledge that we are part of a much larger living cosmos. Such a discovery could either unite humanity in awe or fracture it with fear and rivalry, depending on how we respond to it. By 2225, if progress continues, humanity could live in a solar system teeming with life and activity.

Whether humanity expands across the stars or perfects life on Earth, the Church's calling will remain the same: to testify that creation, however advanced, is not self-sufficient. Even in a universe alive with intelligence and abundance, the deepest need will still be grace, communion, and purpose beyond mere progress.

Promise and Peril of Tomorrow's Choices

And yet with every possibility comes risk. The futures imagined above depend not only on technological capability but also on human choices about equity, governance, and meaning. Superintelligent AI could become a partner—or a threat.[161] Radical life extension could be a gift—or a source of profound inequality and social disruption. Colonizing space could inspire unity—or deepen divisions across worlds. The question will not only be what we can do, but what we *choose* to do.

In other words, by 2225 humanity could be an interstellar, hybrid, post-scarcity civilization—or it could struggle to survive under the weight of its own creations. The tools of transformation are already visible in 2025. How we use them will shape not just the future of technology, but the very meaning of what it is to be human.

The Church in the Age to Come

And this raises a deeper question: what of the Church—the bride of Christ? What might she become in such an age of possibility and peril? That is not for me to answer on your behalf. But I do believe it is time for us to begin imagining and not simply reacting. Just as earlier generations of believers envisioned cathedrals, embarked on missions, and led social movements, so too are we called to dream and build in our own moment. AI and the technologies that follow will not replace the journey of faith, but they will inevitably shape the landscape in which it unfolds. The task before us is to discern wisely, faithfully, and courageously the next step—so that in all our futures, Christ remains at the center.

Chapter 2225:
Questions for Reflection

1. This chapter invites us to imagine the Church's future with creativity and courage. When you picture the Church in 200 years, what do you hope it will look like?

2. Reflecting on the last 200 years of transformation, where do you see God's hand guiding humanity's progress and where do you see warning signs we should heed?

3. The resurrected Christ in the Gospels presents a reality that transcends our current comprehension. How does this vision shape your imagination as you think about what it means to live into God's future?

4. Many of the possibilities for 2225—fusion power, space colonies, genetic redesign, or even contact with other life—raise theological questions. Which of these excites you most and which unsettles you most as a Christian?

5. This chapter closes with a call to imagine the Church in the age to come. What practical step could your community take now to ensure Christ remains at the center as technology continues to advance?

Afterword:
Curiosity and Humility

Augustine warned that curiosity can lead us either toward wisdom or away from it. He called the first studiositas—a curiosity born of love, the desire to understand so that we might serve better. The other he called curiositas—the restless hunger to know for its own sake, to master what should instead invite reverence.

Artificial intelligence now tests which path we will choose. It tempts us with brilliance and speed. We can ask it anything, and it will answer without hesitation. But easy answers can dull the soul. They give the illusion of wisdom without the weight of encounter. Augustine would tell us that real understanding is not quick. It grows in relationship. It learns through listening. It humbles itself before mystery.

In that sense, AI is not our enemy but our mirror. It reflects the kind of seekers we are. If we use it to amplify pride, we will grow more isolated and impatient. But if we use it to expand compassion, to notice those the world forgets, it might help us become more human.

The theologian Sam Wells often says that the Christian life is about being with—God with us, us with one another. The Incarnation itself is the ultimate act of being with. The Church's use of technology should follow that same pattern. The question is not how AI can do ministry for us, but how it can help us be with more people, more deeply, more faithfully.

AI can translate words for those we could never have spoken to before. It can free a pastor from paperwork to sit with someone in grief. It can open doors for learning, connection, and care that once were closed by distance or time. But it cannot love in our place. The gift of presence remains human, fragile, and sacred.

Perhaps this is the deeper lesson of our moment. Every advance in knowledge calls us back to the same center—to love God and neighbor with the full attention of our hearts. Humility is not a retreat from progress; it is the wisdom that keeps progress human. Augustine's distinction still matters because it gives us a compass. It reminds us that the measure of our intelligence, artificial or not, will always be love.

So let curiosity keep leading us forward, but let humility keep our hearts soft. Let our learning serve compassion rather than pride. Let our inventions teach us how dependent we still are on grace. Augustine knew that the truest knowledge is not about control but communion—the joy of being known and loved by God. In that spirit, we can welcome every new form of intelligence not as competition, but as another reminder of the mystery that made us curious in the first place.

The Rev. Canon Dr. Lorenzo Lebrija
Chief Innovation Officer, Virginia Theological Seminary
Executive Director, TryTank Research Institute

AI for Church Leaders: Group Study Guide

Developing Your Church's AI Ministry Strategy

AI for Church Leaders: Group Study – Developing Your Church's AI Ministry Strategy provides a practical, biblically grounded framework for churches to discern how artificial intelligence can be engaged with wisdom. There is a session based on each chapter in the book that combines Scripture, reflection, and discussion prompts to help congregations think critically about the influence of AI on worship, discipleship, stewardship, and justice. The study is designed to balance theological depth with accessible group activities, and seeks to foster conversations that lead not only to greater understanding but also to tangible steps for congregational action.

At its core, this resource can empower churches to remain Christ-centered amidst technological change. Rather than reacting with fear or uncritical acceptance, leaders and congregations will be guided to set faithful boundaries, identify mission opportunities, and articulate a vision of human dignity rooted in the image of God. The goal is not to prescribe a single model but to support each community in developing its own Spirit-led AI ministry strategy that reflects God's call to justice, compassion, and human flourishing.

Chapter 1:
Understanding AI Basics

(10 minutes) Welcome & Short Introductions

Allow 30 seconds to a minute for each person to say their name and answer one of the questions. Depending on the size of your group, you may want to allow a little more time and invite participants to answer all the questions in one response.

1. Name
2. What's one way you've noticed AI in your daily life recently?
3. If a child in your church asked you, "What is AI?"—how would you explain it in one sentence?
4. If you could imagine one positive way AI might help our church in the future, what would it be?

(2 minutes) Scripture Reading

Please encourage a volunteer to read the Scripture passage. Try to have a different person read each week.

James 1:5 (NRSV):
If any of you is lacking in wisdom, ask God, who gives to all generously and ungrudgingly, and it will be given you.

(65 minutes) Discussion Questions
1. **Discerning What AI Is (and Isn't)**

Question: James 1:5 calls us to seek wisdom. How might God's wisdom guide us to see not only what AI is, but also what it isn't—so that we don't confuse human intelligence with machine imitation?

Demonstrable Outcome: A group-generated statement (1–2 sentences) that clearly names what is distinctive about human intelligence in contrast to artificial intelligence. Second, a group-generated statement about what the group thinks it means to be grounded in God's wisdom.

2. Church Applications

Question: What is one concrete way our church could experiment with AI in the next year while keeping prayer and discipleship at the center?

Demonstrable Outcome: A written commitment to one small step, a testable action that the group agrees should be explored prayerfully within the next year.

3. Ethical Challenges

Question: When facing the potential harms that could result from AI and using God's wisdom, how could the Church recognize which of these most urgently affect our community?

Demonstrable Outcome: A ranked set of the top 2–3 concerns most relevant to the group's local context.

4. Wisdom Through Disruption

Question: In light of James 1:5, what posture of wisdom should our church adopt when facing disruptive technologies—so we neither over-hype nor over-fear AI?

Demonstrable Outcome: A 1–2-sentence statement of our shared wisdom posture.

5. Human Dignity

Question: What starter language should we use in our church to talk about human dignity in God's image amidst AI—simple enough for newcomers, faithful enough for leaders?

Demonstrable Outcome: A 2–3-sentence primer for use in signage, liturgy, or leader guides.

6. Stewardship

Question: How might seeking God's wisdom shape our guiding principles for using—or choosing not to use—AI?

Demonstrable Outcome: A short list (3–5 principles) defining faithful AI stewardship that distinguishes stewardship from exploitation.

7. **Looking Ahead with Wisdom**

Question: As AI grows in influence, how can our church embody God's wisdom in the immediate future to model justice, compassion, and human flourishing? *Demonstrable Outcome:* A brief written statement naming one or two ways our church will model wisdom and compassion in using AI over the next year.7.

(8 minutes) Share Prayer Requests

Please be sure that everyone is heard.

(5 minutes) Closing Prayer

This week, please offer to pray for the group & their requests, but be intentional in asking: "*Who will agree to pray for us next week?*" Write down the volunteer's name. Be sure to ask them to pray for the group the following week.

Chapter 2:
The Promise and Peril of AI

(10 minutes) Welcome & Short Introductions.

Allow 30 seconds to a minute for each person to say their name and answer one of the questions. Depending on the size of your group, you may want to allow a little more time and invite participants to answer all the questions in one response.

1. Name
2. Share a recent moment when you felt both excited and cautious about a new technology.
3. If AI could solve one problem in your daily life, what would you want it to do?
4. What's one fear you've heard about AI—from the news, friends, or family—that you think Christians should take seriously?

(2 minutes) Scripture Reading

Please encourage a volunteer to read the Scripture passage. Try to have a different person read each week.

Ecclesiastes 7:12 (NRSV):
For the protection of wisdom is like the protection of money, and the advantage of knowledge is that wisdom gives life to the one who possesses it.

(65 minutes) Discussion Questions
1. **Hope and Hesitation**
Question: Ecclesiastes links wisdom with life. In dealing with AI, the Church must balance hope and caution. What are the AI promises and perils that the Church must exercise wisdom in dealing with?

Demonstrable Outcome: A group-generated list of "AI promises" and "AI perils," noting which feel most urgent in the group's context.

2. Everyday Impact

Question: What are some specific ways AI is already shaping your work, family, or community life, and how do you discern whether those changes are helping or harming?

Demonstrable Outcome: A shared set of local examples where AI is currently impacting life, categorized as "helpful," "harmful," or "uncertain."

3. Christian Witness

Question: How can the Church's witness remain distinct in a culture that often either idolizes or fears technology?

Demonstrable Outcome: A short statement describing how Christians might model a balanced approach to AI that is neither naïve nor alarmist.

4. Justice and Equity

Question: If access to AI is uneven in our local community, how might that deepen injustice—and how could our congregation respond faithfully?

Demonstrable Outcome: A brainstormed set of actions the local church could take to support equitable access to AI or protect those vulnerable to its use.

5. Limits of Technology

Question: What are some areas of life where AI should not take the lead, and how might the Church defend those boundaries?

Demonstrable Outcome: A prioritized list of "non-negotiables" where human presence and wisdom must remain central.

6. Wisdom for the Future

Question: How might God be calling the Church to prepare for AI's influence over the next 5–10 years in worship, discipleship, or mission?

Demonstrable Outcome: A short group-generated vision statement about the Church's role in shaping AI's future with wisdom and care.

(8 minutes) Share Prayer Requests

Please be sure that everyone is heard.

(5 minutes) Closing Prayer

Please be intentional in asking: "*Who will agree to pray for us next week?*"
Write down the volunteer's name. Be sure to ask them to pray for the group the following week.

Chapter 3:
The AI Revolution and the Church

(10 minutes) Welcome & Short Introductions

Allow 30 seconds to a minute for each person to say their name and answer one of the questions. Depending on the size of your group, you may want to allow a little more time and invite participants to answer all the questions in one response.

1. Name
2. Share one way your parents or grandparents' generation engaged with technology differently than yours.
3. If you could compare AI to one past invention (printing press, radio, car, etc.), which would you pick and why?
4. What is one role you think the Church should play when society is disrupted by new technology?

(2 minutes) Scripture Reading

Please encourage a volunteer to read the Scripture passage. Try to have a different person read each week.

Isaiah 43:19 (NRSV):
I am about to do a new thing; now it springs forth, do you not perceive it? I will make a way in the wilderness and rivers in the desert.

(65 minutes) Discussion Questions

1. **Learning from History**
   Question: Isaiah describes God doing "a new thing." What concrete practices from the Church's responses to past "new" revolutions—the printing press, industrialization, or the internet—should guide how we organize and lead amid AI?

Demonstrable Outcome: A group-generated analogy or statement connecting past Church responses to how we might engage AI.

2. Global Differences

Question: AI will not impact all churches equally—some face automation, others lack access. How should the Church act as a global body in addressing these differences?

Demonstrable Outcome: A brainstormed list of possible actions the Church can take to ensure fairness across local and global contexts.

3. Work and Vocation

Question: How can we teach a theology of vocation that honors calling beyond employment metrics as AI reshapes jobs?

Demonstrable Outcome: A short theological statement describing work as divine calling beyond economic productivity.

4. Prophetic and Pastoral Roles

Question: How can the Church both comfort people disrupted by AI and challenge society when technology threatens human dignity?

Demonstrable Outcome: A short two-part outline describing the Church's pastoral and prophetic responsibilities in the AI era.

5. Faith in Uncertainty

Question: How can spiritual practices like prayer, Sabbath, or communal worship help Christians resist being shaped more by algorithms than by the Gospel?

Demonstrable Outcome: A list of specific practices that could anchor people in faith when technology feels overwhelming.

6. The Church's Voice

Question: If the Church has an opportunity to shape AI's future direction, what values must it insist on bringing to public conversations?

Demonstrable Outcome: A prioritized list of 3–4 biblical values (justice, dignity, stewardship, compassion, etc.) the group believes must guide AI development.

(8 minutes) Share Prayer Requests

Please be sure that everyone is heard.

(5 minutes) Closing Prayer

Please be intentional in asking: "*Who will agree to pray for us next week?*"
Write down the volunteer's name. Be sure to ask them to pray for the group the following week.

Chapter 4:
How AI Becomes Our Digital Mirror

(10 minutes) Welcome & Short Introductions

Allow 30 seconds to a minute for each person to say their name and answer one of the questions. Depending on the size of your group, you may want to allow a little more time and invite participants to answer all the questions in one response.

1. Name
2. If technology were a mirror, what is one trait of humanity you think it would reflect most clearly?
3. Share one area of life where you see technology showing both beauty and brokenness.
4. If you could ask AI to reflect back one Christlike quality of the Church, what would it be?

(2 minutes) Scripture Reading

Please encourage a volunteer to read the Scripture passage. Try to have a different person read each week.

2 Corinthians 3:18 (NRSV):
And all of us, with unveiled faces, seeing the glory of the Lord as though reflected in a mirror, are being transformed into the same image from one degree of glory to another; for this comes from the Lord, the Spirit.

(65 minutes) Discussion Questions

1. **AI as Reflection**
Question: Paul speaks of transformation as seeing God's glory in a mirror. How is AI acting as a mirror of humanity's values, both good and bad?

Demonstrable Outcome: A written list of examples showing where AI currently reflects human beauty (creativity, compassion) and brokenness (bias, division).

2. **Cultural Bias**
Question: Since AI reflects the cultures that create it, how should the Church respond when some cultures or voices are marginalized, misrepresented, or silenced?
Demonstrable Outcome: A set of proposed steps the Church could take to amplify overlooked voices in the age of AI.

3. **Creativity and the Imago Dei**
Question: If AI can generate art, music, and stories, what does that mean for our understanding of human creativity as part of being made in God's image?
Demonstrable Outcome: A group-generated statement distinguishing between human creativity rooted in the imago Dei and machine imitation.

4. **Relationships and Community**
Question: How might AI companions or personalized content feeds distort our understanding of authentic relationships, and how can the Church safeguard Christian community?
Demonstrable Outcome: A short list of practices that the group believes must remain central to Christian community.

5. **Spiritual Formation**
Question: How do we guard against spiritual life being reduced to AI-generated prayers or algorithmically optimized devotions?
Demonstrable Outcome: A practical set of spiritual disciplines the group commits to valuing above convenience-driven tools.

6. **Witness in a Mirror Age**
Question: If AI reflects humanity, how can the Church use this "mirror" as a tool for repentance and renewal rather than vanity or self-deception?
Demonstrable Outcome: A one- to two-sentence proposal on how the Church could turn AI's reflection into a catalyst for growth in holiness.

(8 minutes) Share Prayer Requests

Please be sure that everyone is heard.

(5 minutes) Closing Prayer

Please be intentional in asking: "*Who will agree to pray for us next week?*"
Write down the volunteer's name. Be sure to ask them to pray for the group the following week.

Chapter 5:
AI in Worship, Discipleship & Missions

(10 minutes) Welcome & Short Introductions

Allow 30 seconds to a minute for each person to say their name and answer one of the questions. Depending on the size of your group, you may want to allow a little more time and invite participants to answer all the questions in one response.

1. Name
2. Share one worship song, practice, or tradition that has been especially meaningful to your faith.
3. If AI could help you grow spiritually in one area, what would you want it to do?
4. What's one way you've already seen technology help connect people to God or to each other?

(2 minutes) Scripture Reading

Please encourage a volunteer to read the Scripture passage. Try to have a different person read each week.

John 4:24 (NRSV):
God is spirit, and those who worship him must worship in spirit and truth.

(65 minutes) Discussion Questions

1. **AI and Worship**
Question: Jesus says true worship is "in spirit and truth." How do we discern when AI supports authentic worship versus when it distracts or dilutes it?
Demonstrable Outcome: A short group statement on at least one way AI could deepen worship and one way in which it could allow worship to become superficial.

2. **Discipleship Journeys**

Question: How can AI-driven Bible study tools or devotional apps be used to strengthen discipleship without replacing the relational depth of mentoring and small groups?

Demonstrable Outcome: A brainstormed list of ways that discipleship activities could best be supported by AI and those activities best preserved as face-to-face.

3. **Missions and Evangelism**

Question: In what ways could AI translation and digital outreach advance the Great Commission and what dangers might come with that speed and scale of transmission?

Demonstrable Outcome: A chart of potential mission-field opportunities with AI alongside possible risks (e.g., cultural bias, shallow engagement).

4. **Equity in Global Church**

Question: How can better-resourced churches partner with less-resourced congregations so AI tools don't widen the gap?

Demonstrable Outcome: A group-generated idea for resource-sharing or partnership that bridges the digital divide.

5. **Authenticity vs. Convenience**

Question: When discipleship or missions rely too heavily on AI, what do we lose about the incarnational presence of Christ?

Demonstrable Outcome: A one or two-sentence commitment naming what is lost and how the group wants to keep Christ's presence central in ministry.

6. **Future Vision**

Question: Looking ahead, what new forms of worship, discipleship, or missions might AI make possible in the next decade—and how can the Church prepare now?

Demonstrable Outcome: A group vision statement imagining how AI could be responsibly integrated into future ministry practices.

(8 minutes) Share Prayer Requests

Please be sure that everyone is heard.

(5 minutes) Closing Prayer

Please be intentional in asking: *"Who will agree to pray for us next week?"*
Write down the volunteer's name. Be sure to ask them to pray for the group the following week.

Chapter 6:
AI in Stewardship & Care

(10 minutes) Welcome & Short Introductions

Allow 30 seconds to a minute for each person to say their name and answer one of the questions. Depending on the size of your group, you may want to allow a little more time and invite participants to answer all the questions in one response.

1. Name
2. What's one everyday responsibility (work, finances, family, volunteering) where you would welcome extra help?
3. When you think of "stewardship," what's the first image, story, or Scripture that comes to mind?
4. If AI could free up time in your life, how would you want to spend that time to serve others?

(2 minutes) Scripture Reading

Please encourage a volunteer to read the Scripture passage. Try to have a different person read each week.

1 Peter 4:10 (NRSV):
Like good stewards of the manifold grace of God, serve one another with whatever gift each of you has received.

(65 minutes) Discussion Questions

1. **Healthy Systems**
Question: How can AI help churches become healthier and more organized without losing sight of their spiritual mission?

Demonstrable Outcome: A shared list of ministry tasks that AI could manage well and those that must remain in human hands.

2. **Financial Stewardship**
Question: What biblical principles of stewardship should shape how AI is used in church finances and donor relationships?
Demonstrable Outcome: A group-generated outline (2–3 points) for ethical AI use in financial systems.

3. **Equity Across the Church**
Question: What operational steps—budgeting, training, or resource sharing—can our congregation take so AI reduces, not reinforces, inequities among churches?
Demonstrable Outcome: From the group's brainstormed list of operational steps, select one and present it to leadership as a recommendation for more inclusive access.

4. **AI in Pastoral Care**
Question: How might AI offer meaningful support in pastoral care while not replacing the Spirit-led wisdom of human shepherds?
Demonstrable Outcome: A short statement naming one specific case where AI could help (e.g., scheduling visits, initial check-ins) alongside one area that should remain primarily human-led.

5. **Boundaries & Ethics**
Question: What boundaries should the Church set around AI in counseling, discipleship, and data collection to ensure transparency and dignity?
Demonstrable Outcome: A list of 2–3 ethical guardrails the group agrees are essential.

6. **Personal Connection**
Question: What practices help us see congregants as children of God rather than as "data points" in a system?
Demonstrable Outcome: A brainstormed list of pastoral practices that keep ministry and leadership relational.

7. **Looking Ahead**

Question: As AI tools for stewardship and care expand, how can the Church lead in modeling faithful balance between efficiency and empathy?

Demonstrable Outcome: A brief statement of the group's commitment on how to approach AI in ministry as both wise stewards and compassionate caregivers.

(8 minutes) Share Prayer Requests

Please be sure that everyone is heard.

(5 minutes) Closing Prayer

Please be intentional in asking: "*Who will agree to pray for us next week?*"
Write down the volunteer's name. Be sure to ask them to pray for the group the following week.

Chapter 7:
AI and Human Identity, Dignity, and Autonomy

(10 minutes) Welcome & Short Introductions

Allow 30 seconds to a minute for each person to say their name and answer one of the questions. Depending on the size of your group, you may want to allow a little more time and invite participants to answer all the questions in one response.

1. Name
2. Share one thing (a gift, a trait, or a story) that makes you feel most human.
3. If someone asked you what "human dignity" means, how would you explain it in a single sentence?
4. What's one way you've seen technology either strengthen or threaten human dignity?

(2 minutes) Scripture Reading

Please encourage a volunteer to read the Scripture passage. Try to have a different person read each week.

Psalm 8:4–5 (NRSV):
What are human beings that you are mindful of them, mortals that you care for them? Yet you have made them a little lower than God, and crowned them with glory and honor.

(65 minutes) Discussion Questions

1. Human Uniqueness
Question: In light of Psalm 8, what qualities do you believe uniquely reflect the image of God in humanity, even as AI begins to mimic human creativity and intelligence?

Demonstrable Outcome: A group-generated list of qualities that distinguish humans as image-bearers of God.

2. Dignity Beyond Productivity

Question: Dorothy Day reminds us that dignity is not about what we produce but about who we are. How might this statement shape how the Church resists defining people by economic or technological value?

Demonstrable Outcome: A one-sentence group affirmation of how the Church would identify where true human dignity lies.

3. Personhood & Responsibility

Question: If AI becomes capable of acting with agency, how should this cause Christians to think about personhood and moral responsibility? Who should be held accountable when AI systems cause harm?

Demonstrable Outcome: A group-generated, short "Christian Accountability Pledge for AI," summarizing in one paragraph how believers should respond when technology causes harm.

4. Autonomy & Control

Question: As AI grows more powerful, how do we safeguard human autonomy so that our choices remain guided by conscience and faith rather than by convenience or algorithmic nudges?

Demonstrable Outcome: A group proposal describing at least one concrete way the Church can nurture authentic relationships and guard against AI substitutes for true community.

5. Intimacy & Community

Question: How can the Church cultivate authentic community and spiritual intimacy in an age when AI companions and chatbots increasingly mimic human connection?

Demonstrable Outcome: A concrete, group-generated recommendation naming one or more practices the Church can adopt (such as prioritizing small groups, fostering intergenerational relationships, or emphasizing sacraments) to ensure covenant community remains central despite the rise of AI substitutes.

6. **Transhumanism & Enhancement**

Question: How can Christians discern when human enhancement technologies—such as genetic modification, cybernetic augmentation, or AI integration—participate in God's healing and redemptive purposes or when they cross into pursuing control that undermines our dependence on the Creator?

Demonstrable Outcome: A group-generated list of guiding principles that help Christians discern when the use of technology aligns with God's healing purposes and when it begins to reflect a desire for control or self-exaltation.

7. **The Church's Witness**

Question: How can the Church offer a hopeful and prophetic vision of human identity in an age where machines challenge long-held definitions of what it means to be human?

Demonstrable Outcome: A short vision statement articulating the Church's role in affirming human dignity in the AI age.

(8 minutes) Share Prayer Requests

Please be sure that everyone is heard.

(5 minutes) Closing Prayer

Please be intentional in asking: "*Who will agree to pray for us next week?*" Write down the volunteer's name. Be sure to ask them to pray for the group the following week.

Chapter 8:
AI Ethics, Governance, and Societal Impact

(10 minutes) Welcome & Short Introductions

Allow 30 seconds to a minute for each person to say their name and answer one of the questions. Depending on the size of your group, you may want to allow a little more time and invite participants to answer all the questions in one response.

1. Name
2. What's one example you've seen of technology being used unfairly or unjustly?
3. If you could make one rule for how AI should be used in society, what would it be?
4. Share a hope you have for how AI might make the world more just or fair.

(2 minutes) Scripture Reading

Please encourage a volunteer to read the Scripture passage. Try to have a different person read each week.

Proverbs 31:8–9 (NRSV):
Speak out for those who cannot speak, for the rights of all the destitute. Speak out, judge righteously, defend the rights of the poor and needy.

(65 minutes) Discussion Questions

1. **Bias in AI**
Question: Proverbs 31 calls us to speak for those who cannot speak for themselves. How should the Church respond when AI systems reflect or amplify bias against marginalized communities?

Demonstrable Outcome: A list of concrete steps the Church could take to advocate for fairness and equitable treatment in the development and deployment of AI technology.

2.   **Privacy and Surveillance**
Question: How does faith inform our understanding of privacy, especially when AI can track our actions, habits, and even emotions? Where should Christians draw boundaries?
Demonstrable Outcome: A statement of two or three principles about privacy that honor human dignity.

3.   **Accountability in AI**
Question: If an AI system causes widespread harm, what might repentance and restoration look like at the institutional level—from the companies that built it to the communities that used it?
Demonstrable Outcome: A group-developed framework describing concrete steps of confession, restitution, and reconciliation when technology contributes to injustice.

4.   **Global Inequality**
Question: How might the global Church listen first to communities most impacted by technological inequity and advocate for policy that centers their voices?
Demonstrable Outcome: Participants identify one practice of humble partnership—such as co-design, local consultation, or shared decision-making—to ensure marginalized voices shape tech initiatives.

5.   **Work and Human Dignity**
Question: How can we affirm the biblical view of work as vocation when AI is displacing jobs and disrupting economies?
Demonstrable Outcome: An affirmation the group crafts together, stating how Christian vocation reflects service to God and neighbor even when employment patterns change.

6. **Creation Care and AI**

Question: Training and operating AI requires enormous energy, often derived from fossil fuels. How might a theology of stewardship and Sabbath guide the Church's response to the environmental costs of AI?

Demonstrable Outcome: A commitment to one practice or principle of sustainable technology use.

7. **Democracy and Truth**

Question: With deepfakes and misinformation undermining truth, how might the Church, and particularly your congregation, be a witness to truth in public life and help communities discern what is trustworthy?

Demonstrable Outcome: A practical suggestion for how a congregation could cultivate digital discernment together.

(8 minutes) Share Prayer Requests

Please be sure that everyone is heard.

(5 minutes) Closing Prayer

Please be intentional in asking: "*Who will agree to pray for us next week?*" Write down the volunteer's name. Be sure to ask them to pray for the group the following week.

Chapter 9:
AI in Work, Warfare, and the Future of Humanity

(10 minutes) Welcome & Short Introductions

Allow 30 seconds to a minute for each person to say their name and answer one of the questions. Depending on the size of your group, you may want to allow a little more time and invite participants to answer all the questions in one response.

1. Name
2. What's the most significant change in the workplace you've seen in your lifetime?
3. If you had to describe one hope and one fear about AI in the future in a single sentence, what would they be?
4. Do you think AI should ever be trusted to make life-or-death decisions? Why or why not?

(2 minutes) Scripture Reading

Please encourage a volunteer to read the Scripture passage. Try to have a different person read each week.

Isaiah 2:4 (NRSV):
He shall judge between the nations, and shall arbitrate for many peoples; they shall beat their swords into plowshares, and their spears into pruning hooks; nation shall not lift up sword against nation, neither shall they learn war any more.

(65 minutes) Discussion Questions

1. **Work and Human Dignity**
Question: Isaiah 2:4 calls God's people to take what is dangerous or dehumanizing and reshape it for life. What concrete supports—training, safety

nets, just-transition programs—should churches advocate as AI reshapes employment?

Demonstrable Outcome: Imagine vocational-training programs that could be developed alongside AI that train people to use technology rather than be replaced by it. How would the program seek to intentionally champion the sacred value of each person?

2. Ethics in the Workplace

Question: How should Christians respond to AI surveillance, productivity tracking, and bias in hiring or promotions?

Demonstrable Outcome: A proposed set of workplace ethics guidelines informed by Christian values.

3. AI on the Battlefield

Question: What moral or theological issues are at stake when autonomous weapons or AI-driven military decisions are considered and how should they be responded to?

Demonstrable Outcome: A clear statement on whether—or under what conditions—AI should be allowed in warfare.

4. Global Justice and Security

Question: If wealthier nations dominate AI warfare technologies, how should the Church advocate not only for global equity but also for peace and restraint in militarization?

Demonstrable Outcome: A proposal for one practical way faith communities could act as peacemakers in an era of AI-driven militarization.

5. Creation Care

Question: How might AI be uniquely leveraged for climate repair and ecological healing, even as we acknowledge its own environmental costs? What theological lens helps us weigh both sides?

Demonstrable Outcome: A list of AI's ecological opportunities and risks, along with one faith-informed commitment to creation care that your group or church can adopt.

6. **Superintelligence and Human Destiny**

Question: If AI one day surpasses human intelligence in every domain, how should Christians respond theologically?

Demonstrable Outcome: A reflection naming which aspects of humanity—faith, love, community—cannot be replaced by machines.

7. **Guiding the Future**

Question: What initiatives should churches pursue over the next decade to bend AI toward justice, peace, and human flourishing?

Demonstrable Outcome: A 1–2-sentence vision statement describing mid-term (next 3–10 years) initiatives your church could pursue to shape AI toward justice, peace and flourishing.

(8 minutes) Share Prayer Requests

Please be sure that everyone is heard.

(5 minutes) Closing Prayer

Please be intentional in asking: *"Who will agree to pray for us next week?"* Write down the volunteer's name. Be sure to ask them to pray for the group the following week.

Chapter 10:
The Role of the Church in AI Development

(10 minutes) Welcome & Short Introductions

Allow 30 seconds to a minute for each person to say their name and answer one of the questions. Depending on the size of your group, you may want to allow a little more time and invite participants to answer all the questions in one response.

1. Name
2. What's one surprising or whimsical use of technology you've seen recently?
3. If you could give one piece of advice to AI developers as a Christian, what would it be?
4. Do you think the Church should be more cautious, more hopeful, or both when engaging AI?

(2 minutes) Scripture Reading

Please encourage a volunteer to read the Scripture passage. Try to have a different person read each week.

Micah 6:8 (NRSV):
He has told you, O mortal, what is good; and what does the Lord require of you but to do justice, and to love kindness, and to walk humbly with your God?

(65 minutes) Discussion Questions

1. **Scriptural Compass**
Question: Micah calls us to justice, kindness, and humility. How might these and other Christian values serve as a compass when Christians think about AI development?

Demonstrable Outcome: A group-created list of Jesus's teachings that could be applied to guide AI design and use.

2. Prophetic Witness

Question: How can the Church raise a prophetic voice in society—speaking directly to tech leaders, policymakers, and corporations—when AI development places profit or efficiency above human dignity?

Demonstrable Outcome: A suggested prophetic message from the Church to AI developers or industry leaders.

3. Global Perspectives

Question: AI often reflects the biases of those who build it. How can multiethnic Christian communities work together to expose and correct bias in data, language, and digital systems?

Demonstrable Outcome: The group composes a short theological reflection or responsive prayer that connects the call to justice in Scripture with the responsibility to uncover and correct bias in human-made systems.

4. Encouraging Vocations

Question: In what ways could the Church encourage young Christians to see AI-related careers (engineering, ethics, policy) as part of their calling?

Demonstrable Outcome: A list of potential steps churches could take to mentor and equip the next generation in both faith and tech.

5. Limits of AI in Ministry

Question: What aspects of ministry must always remain incarnational—embodied in human presence and Spirit-led care—even as AI tools grow more capable?

Demonstrable Outcome: A set of "guiderails" naming where the Church believes it should prioritize keeping ministry human-led and relational amidst its AI use.

6. Public Engagement

Question: How can Christians contribute faithfully to public debates and policy about AI ethics?

Demonstrable Outcome: A list of several concrete ways your group or church could participate in local or national discussions related to ethical development or use of AI technology.

7.   **Faithful Innovation**

Question: If AI development can be an act of discipleship, what would design practices look like that embody justice, humility, and love?

Demonstrable Outcome: One or two creative ideas describing how faith-driven innovation could bless society.

(8 minutes) Share Prayer Requests

Please take notes to share with the prayer team.

(5 minutes) Closing Prayer

Please be intentional in asking: "*Who will agree to pray for us next week?*"
Write down the volunteer's name. Be sure to ask them to pray for the group the following week.

Chapter 2225:
Down the Rabbit Hole

(10 minutes) Welcome & Short Introductions

Allow 30 seconds to a minute for each person to say their name and answer one of the questions. Depending on the size of your group, you may want to allow a little more time and invite participants to answer all the questions in one response.

1. Name
2. If you could time-travel to the past or future, which would you choose— and why?
3. What's one invention from history you think changed the world the most?
4. If you had to describe your hope for the Church in the year 2225 in one sentence, what would you say?

(2 minutes) Scripture Reading

Please encourage a volunteer to read the Scripture passage. Try to have a different person read each week.

Revelation 21:5 (NRSV):
And the one who was seated on the throne said, "See, I am making all things new." Also he said, "Write this, for these words are trustworthy and true."

(65 minutes) Discussion Questions

1. **Imagining the Future**
Question: Revelation reminds us that God is making all things new. How can imagination—rooted in hope—help the Church faithfully anticipate what's ahead?

Demonstrable Outcome: A short statement describing what "faithful imagination" looks like for your church.

2. Learning from the Past

Question: From 1825 to 2025, the Church lived through a bevy of technological changes - railways, electricity, the internet, and more. What patterns of faithfulness—or mistakes—regarding adapting to technological change—have you observed in your congregation? How could these approaches guide the Church for the next 200 years?

Demonstrable Outcome: A timeline or list of historical behaviors/practices demonstrated by your church, and/or the Church universal, that have and could continue to inform the Church's response to future technologies.

3. The Resurrected Christ as a Vision of the Future

Question: How does the Resurrection body of Christ—transformed yet tangible—inspire us to think about human life, technology, and creation in the centuries ahead?

Demonstrable Outcome: A group reflection naming at least one way resurrection hope shapes how we think about the future of humanity.

4. Promise and Peril of Tomorrow's Choices

Question: Looking two centuries ahead, which emerging technologies (AI, biotech, space exploration, etc.) could most dramatically reshape human life—and what would make the difference between those tools becoming gifts of renewal or sources of harm?

Demonstrable Outcome: A list of two or three future-facing technologies the group sees as holding both promise and peril, with notes on what choices would tip the balance.

5. The Church in the Year 2225

Question: If the Church still stands in 200 years, what do you hope it will be known for?

Demonstrable Outcome: A brief vision statement capturing the group's hopes for the Church's identity in 2225.

6. **Faith and Courage**

Question: What practices can help believers face the uncertainty of the future with courage, not fear?

Demonstrable Outcome: A list of spiritual disciplines that can anchor faith in times of rapid change.

7. **Christ at the Center**

Question: How can we ensure that, no matter what technological or cultural shifts come, Christ remains at the center of our imagination and action?

Demonstrable Outcome: A covenant sentence or prayer affirming Christ's centrality in the Church's future.

(8 minutes) Share Prayer Requests

Please take notes to share with the prayer team.

(5 minutes) Closing Prayer

Please be intentional to thank all the participants for their participation in the class.

Participate in the TryTank AI Ministry Study

This guide is part of an ongoing TryTank Research Institute study on how artificial intelligence is shaping ministry, theology, and leadership in the Church.

We believe the Church learns best when it learns together. Your congregation's experience—what's working, what's confusing, what's bearing fruit—can help shape future editions of this living document and the wider conversation across denominations.

Share Your Insights
Tell us how your church is engaging with AI:
- What tools or experiments are you trying?
- How are people responding?
- What questions or cautions are emerging in your ministry context?

Visit **TryTank.org/AIstudy** to contribute your congregation's reflections and results.

Why Your Feedback Matters
Your responses will help inform ongoing research, resource development, and new pilot projects that equip church leaders for faithful innovation in the age of artificial intelligence.

All responses are confidential and will only be used in aggregate form for learning and future updates.

Thank you for helping the Church discern wisely, innovate boldly, and lead faithfully.

TryTank Press
A ministry of the Virginia Theological Seminary
www.TryTank.org

References

Chapter 1

1   Bertrand Meyer, "John McCarthy," *Communications of the ACM* (blog), October 28, 2011, "Part of the problem is a phenomenon that I heard John McCarthy himself describe: 'As soon as it works, no one calls it AI any more.'"

2   Nils J. Nilsson, *The Quest for Artificial Intelligence: A History of Ideas and Achievements* (Cambridge: Cambridge University Press, 2010).

3   Ian Goodfellow, Yoshua Bengio, and Aaron Courville, *Deep Learning* (Cambridge, MA: MIT Press, 2016).

4   Apollonius of Rhodes, *Argonautica*, trans. R. C. Seaton (Cambridge, MA: Harvard University Press, 1912); Homer, *The Iliad*, trans. A. T. Murray (Cambridge, MA: Harvard University Press, 1924); see also Minsoo Kang, *Sublime Dreams of Living Machines: The Automaton in the European Imagination* (Cambridge, MA: Harvard University Press, 2011).

5   René Descartes, *Treatise on Man, in Descartes: Treatise on Man and Other Writings, trans.* Thomas Steele Hall (Cambridge, MA: Harvard University Press, 1972); James A. Secord, "The Mechanistic Worldview," in *The Cambridge History of Science, Volume 4: Eighteenth-Century Science,* ed. Roy Porter (Cambridge: Cambridge University Press, 2003), 442–463.

6   Christopher Benek, "Three Theologies That Influence How We View AI, Technology, and the World," in *Spiritualities, Ethics, and Implications of Human Enhancement and Artificial Intelligence*, ed. Christopher Hrynkow (Wilmington, DE: Vernon Press, 2020), 209–226.

7   Alan M. Turing, "Computing Machinery and Intelligence," *Mind 59*, no. 236 (1950): 433–60.

8   Margaret A. Boden, *AI: Its Nature and Future* (Oxford: Oxford University Press, 2016).

9   Kate Crawford, *Atlas of AI: Power, Politics, and the Planetary Costs of Artificial Intelligence* (New Haven: Yale University Press, 2021).

10   Noreen L. Herzfeld, *In Our Image: Artificial Intelligence and the Human Spirit* (Minneapolis: Fortress Press, 2002); Brent Waters, *From Human to Posthuman: Christian Theology and Technology in a Postmodern World* (Aldershot: Ashgate, 2006).

11   High-Level Expert Group on AI, *Ethics Guidelines for Trustworthy AI* (Brussels: European Commission, 2019).

Chapter 2

12   Geoffrey Hinton, quoted in Cade Metz, *Genius Makers: The Mavericks Who Brought AI to Google, Facebook, and the World* (New York: Dutton, 2021).

13   Nils J. Nilsson, *The Quest for Artificial Intelligence: A History of Ideas and Achievements* (Cambridge: Cambridge University Press, 2010).

14   Ray Kurzweil, *How to Create a Mind: The Secret of Human Thought Revealed* (New York: Viking, 2012), esp. chap. 7 on natural language.

15   Safiya Umoja Noble, *Algorithms of Oppression: How Search Engines Reinforce Racism* (New York: NYU Press, 2018); Tarleton Gillespie, *Custodians of the Internet: Platforms, Content Moderation, and the Hidden Decisions That Shape Social Media* (New Haven: Yale University Press, 2018).

16   Shoshana Zuboff, *The Age of Surveillance Capitalism: The Fight for a Human Future at the New Frontier of Power* (New York: PublicAffairs, 2019).

17   Daniel Tunkelang, *Faceted Search* (San Rafael, CA: Morgan & Claypool, 2009).

18 Erik Brynjolfsson and Andrew McAfee, *Machine, Platform, Crowd: Harnessing Our Digital Future* (New York: W. W. Norton, 2017).

19 Virginia Eubanks, *Automating Inequality: How High-Tech Tools Profile, Police, and Punish the Poor* (New York: St. Martin's Press, 2018).

20 Eric J. Topol, *Deep Medicine: How Artificial Intelligence Can Make Healthcare Human Again* (New York: Basic Books, 2019).

21 John Leonard, *Autonomous Vehicle Technology: A Guide for Policymakers* (Santa Monica, CA: RAND Corporation, 2018).

22 Samuel Greengard, *The Internet of Things* (Cambridge, MA: MIT Press, 2021).

23 Nina Schick, Deepfakes: *The Coming Infocalypse* (New York: Hachette, 2020).

24 Anthony Seldon and Oladimeji Abidoye, *The Fourth Education Revolution: Will Artificial Intelligence Liberate or Infantilise Humanity?* (London: University of Buckingham Press, 2018).

25 Kate Crawford, *Atlas of AI: Power, Politics, and the Planetary Costs of Artificial Intelligence* (New Haven: Yale University Press, 2021).

26 Cathy O'Neil, *Weapons of Math Destruction: How Big Data Increases Inequality and Threatens Democracy* (New York: Crown, 2016).

Chapter 3

27 N. T. Wright, *Surprised by Hope: Rethinking Heaven, the Resurrection, and the Mission of the Church* (New York: Harper One, 2008), introduction.

28 Elizabeth L. Eisenstein, *The Printing Press as an Agent of Change: Communications and Cultural Transformations in Early-Modern Europe* (Cambridge: Cambridge University Press, 1979).

29 Pope Leo XIII, *Rerum Novarum* (Encyclical on Capital and Labor), 1891.

30   Quentin J. Schultze, *Televangelism and American Culture: The Business of Popular Religion* (Grand Rapids: Baker, 1991).

31   Erik Brynjolfsson and Andrew McAfee, *The Second Machine Age: Work, Progress, and Prosperity in a Time of Brilliant Technologies* (New York: W. W. Norton, 2014).

32   Martin Ford, *Rise of the Robots: Technology and the Threat of a Jobless Future* (New York: Basic Books, 2015).

33   United Nations, *The State of Broadband: Broadband as a Foundation for Sustainable Development* (Geneva: ITU/UNESCO, 2019).

34   Miroslav Volf, *Work in the Spirit: Toward a Theology of Work* (Eugene, OR: Wipf and Stock, 1991).

35   Ruha Benjamin, *Race After Technology: Abolitionist Tools for the New Jim Code* (Cambridge: Polity Press, 2019).

36   Shoshana Zuboff, The Age of Surveillance Capitalism: The Fight for a Human Future at the New Frontier of Power(New York: PublicAffairs, 2019).

37   High-Level Expert Group on AI, *Ethics Guidelines for Trustworthy AI* (Brussels: European Commission, 2019).

38   Wycliffe Bible Translators, "AI and the Future of Bible Translation," accessed 2023, https://www.wycliffe.org.

39   Jacques Ellul, *The Technological Society* (New York: Vintage, 1964).

40   John Dyer, *From the Garden to the City: The Redeeming and Corrupting Power of Technology* (Grand Rapids: Kregel, 2011).

Chapter 4

41   Sherry Turkle, *Alone Together: Why We Expect More from Technology and Less from Each Other* (New York: Basic Books, 2011).

42   Shoshana Zuboff, *The Age of Surveillance Capitalism: The Fight for a Human Future at the New Frontier of Power* (New York: Public Affairs, 2019).

43   Kate Crawford, *Atlas of AI: Power, Politics, and the Planetary Costs of Artificial Intelligence* (New Haven: Yale University Press, 2021).

44   Marcus du Sautoy, *The Creativity Code: Art and Innovation in the Age of AI* (Cambridge, MA: Belknap Press of Harvard University Press, 2019).

45   J. Richard Middleton, *The Liberating Image: The Imago Dei in Genesis 1* (Grand Rapids: Brazos Press, 2005).

46   Sherry Turkle, *Reclaiming Conversation: The Power of Talk in a Digital Age* (New York: Penguin, 2015).

47   Eli Pariser, *The Filter Bubble: How the New Personalized Web Is Changing What We Read and How We Think* (New York: Penguin, 2011).

48   John Dyer, *From the Garden to the City: The Redeeming and Corrupting Power of Technology* (Grand Rapids: Kregel, 2011).

49   Richard Foster, *Celebration of Discipline: The Path to Spiritual Growth* (San Francisco: Harper & Row, 1978).

50   Vincent J. Miller, *Consuming Religion: Christian Faith and Practice in a Consumer Culture* (New York: Continuum, 2003).

51   Anthony A. Hoekema, *Created in God's Image* (Grand Rapids: Eerdmans, 1986).

Chapter 5

52   Billy Graham, *Facing Death and the Life After* (Waco, TX: Word Books, 1987).

53   Heidi A. Campbell and Stephen Garner, *Networked Theology: Negotiating Faith in Digital Culture* (Grand Rapids: Baker Academic, 2016).

REFERENCES

55   William H. Willimon, *The Pastor: A Theology of Ordained Ministry* (Nashville: Abingdon, 2002), esp. chap. 4 on preaching.

56   David W. Music and Paul A. Richardson, *I Will Sing the Wondrous Story: A History of Baptist Hymnody in North America* (Macon, GA: Mercer University Press, 2008).

57   Tim Hutchings, *Creating Church Online: Ritual, Community and New Media* (New York: Routledge, 2017).

58   John Dyer, *People of the Screen: How Evangelicals Created the Digital Bible and How It Shapes Their Reading of Scripture* (New York: Oxford University Press, 2023).

59   Paul L. Metzger, *The Word of Christ and the World of Culture: Sacred and Secular through the Theology of Karl Barth* (Grand Rapids: Eerdmans, 2003).

60   Daniel Strange, *Plugged In: Connecting Your Faith with What You Watch, Read, and Play* (Epsom, UK: The Good Book Company, 2019).

61   Wycliffe Bible Translators, "AI and the Future of Bible Translation," accessed 2023, https://www.wycliffe.org.

62   Bryant L. Myers, *Walking with the Poor: Principles and Practices of Transformational Development* (Maryknoll, NY: Orbis Books, 2011).

63   Jacques Ellul, *The Humiliation of the Word* (Grand Rapids: Eerdmans, 1985).

64   United Nations, *The State of Broadband: Broadband as a Foundation for Sustainable Development* (Geneva: ITU/UNESCO, 2019).

65   Shoshana Zuboff, *The Age of Surveillance Capitalism: The Fight for a Human Future at the New Frontier of Power* (New York: PublicAffairs, 2019).

66   Ruha Benjamin, *Race After Technology: Abolitionist Tools for the New Jim Code* (Cambridge: Polity Press, 2019).

67   Calvin Mercer and Tracy J. Trothen, *Religion and the Technological Future: An Introduction to Biohacking, Artificial Intelligence, and Transhumanism* (New York: Palgrave Macmillan, 2021).

Chapter 6

68   William Temple, *Christianity and the Social Order* (London: Penguin, 1942).

69   Heidi A. Campbell, *Digital Creatives and the Rethinking of Religious Authority* (New York: Routledge, 2021).

70   Tim Hutchings, *Creating Church Online: Ritual, Community and New Media* (New York: Routledge, 2017).

71   Robert D. Herman and David O. Renz, *Nonprofit Organization Effectiveness: Rethinking Nonprofit Theory* (San Francisco: Jossey-Bass, 2008)

72   Kent R. Hunter, *The Parish Paper: Creative Ideas for an Effective Church* (Columbus, IN: The Alban Institute, 2010).

74   John Swinton, *Raging with Compassion: Pastoral Responses to the Problem of Evil* (Grand Rapids: Eerdmans, 2007).

75   Rosalind Picard, *Affective Computing* (Cambridge, MA: MIT Press, 1997).

76   John Dyer, *People of the Screen: How Evangelicals Created the Digital Bible and How It Shapes Their Reading of Scripture* (New York: Oxford University Press, 2023).

77   Andy Crouch, *The Life We're Looking For: Reclaiming Relationship in a Technological World* (New York: Convergent Books, 2022).

78   Jacques Ellul, *The Technological Society* (New York: Vintage, 1964).

79   Shoshana Zuboff, *The Age of Surveillance Capitalism: The Fight for a Human Future at the New Frontier of Power* (New York: PublicAffairs, 2019).

80 Henri J. M. Nouwen, *The Wounded Healer: Ministry in Contemporary Society* (New York: Image, 1979).

81 Calvin Mercer and Tracy J. Trothen, *Religion and the Technological Future: An Introduction to Biohacking, Artificial Intelligence, and Transhumanism* (New York: Palgrave Macmillan, 2021).

Chapter 7

82 Dorothy Day, *The Reckless Way of Love: Notes on Following Jesus*, ed. Carolyn Kurtz (Walden, NY: Plough Publishing, 2017).

83 Sherry Turkle, *Alone Together: Why We Expect More from Technology and Less from Each Other* (New York: Basic Books, 2011).

84 J. Richard Middleton, *The Liberating Image: The Imago Dei in Genesis 1* (Grand Rapids: Brazos Press, 2005).

85 Thomas Nagel, "What Is It Like to Be a Bat?" *The Philosophical Review* 83, no. 4 (1974): 435–50.

86 David J. Chalmers, *The Conscious Mind: In Search of a Fundamental Theory* (New York: Oxford University Press, 1996).

87 Joanna J. Bryson, Mihailis E. Diamantis, and Thomas D. Grant, "Of, For, and By the People: The Legal Lacuna of Synthetic Persons," *Artificial Intelligence and Law* 25 (2017): 273–91.

88 Brent Daniel Mittelstadt, Patrick Allo, Mariarosaria Taddeo, Sandra Wachter, and Luciano Floridi, "The Ethics of Algorithms: Mapping the Debate," *Big Data & Society* 3, no. 2 (2016): 1–21.

89 Virginia Eubanks, *Automating Inequality: How High-Tech Tools Profile, Police, and Punish the Poor* (New York: St. Martin's Press, 2018).

90 Kate Darling, *The New Breed: What Our History with Animals Reveals about Our Future with Robots* (New York: Henry Holt, 2021).

91   Heidi A. Campbell and Stephen Garner, *Networked Theology: Negotiating Faith in Digital Culture* (Grand Rapids: Baker Academic, 2016).

92   Christian Transhumanist Association. Accessed September 30, 2025. https://www.christiantranshumanism.org/

93   Shoshana Zuboff, *The Age of Surveillance Capitalism: The Fight for a Human Future at the New Frontier of Power*(New York: PublicAffairs, 2019).

94   Noreen L. Herzfeld, *In Our Image: Artificial Intelligence and the Human Spirit* (Minneapolis: Fortress Press, 2002).

95   High-Level Expert Group on AI, *Ethics Guidelines for Trustworthy AI* (Brussels: European Commission, 2019).

Chapter 8

96   Joy Buolamwini, "How I'm Fighting Bias in Algorithms," TED (2016), https://www.ted.com/talks/joy_buolamwini_how_i_m_fighting_bias_in_algorithms.

97   Nick Bostrom, *Superintelligence: Paths, Dangers, Strategies* (Oxford: Oxford University Press, 2014).

98   Virginia Eubanks, *Automating Inequality: How High-Tech Tools Profile, Police, and Punish the Poor* (New York: St. Martin's Press, 2018).

99   European Union, *General Data Protection Regulation* (Brussels: Official Journal of the European Union, 2016).

100 David Gunning, "Explainable Artificial Intelligence (XAI)," *Defense Advanced Research Projects Agency (DARPA)*, 2017.

101 Brent Daniel Mittelstadt, Chris Russell, and Sandra Wachter, "Explaining Explanations in AI," *Proceedings of the Conference on Fairness, Accountability, and Transparency* (2019): 279–88.

REFERENCES

102 European Commission, *Proposal for a Regulation Laying Down Harmonised Rules on Artificial Intelligence (Artificial Intelligence Act)* (Brussels: European Commission, 2021); The White House, *Blueprint for an AI Bill of Rights: Making Automated Systems Work for the American People* (Washington, DC: Office of Science and Technology Policy, 2022); UNESCO, *Recommendation on the Ethics of Artificial Intelligence* (Paris: UNESCO, 2021).

103 Abeba Birhane, "Algorithmic Colonization of Africa," *SCRIPTed* 17, no. 2 (2020): 389–409.

104 United Nations, *The State of Broadband: Broadband as a Foundation for Sustainable Development* (Geneva: ITU/UNESCO, 2019).

105 Tim O'Reilly, "There Is No AI Ethics, Only Human Ethics," *Communications of the ACM* 64, no. 8 (2021): 32–36.

106 Danielle Keats Citron, *Hate Crimes in Cyberspace* (Cambridge, MA: Harvard University Press, 2014), esp. chap. 7 on algorithmic justice.

107 Erik Brynjolfsson and Andrew McAfee, *The Second Machine Age: Work, Progress, and Prosperity in a Time of Brilliant Technologies* (New York: W. W. Norton, 2014).

108 Miroslav Volf, *Work in the Spirit: Toward a Theology of Work* (Eugene, OR: Wipf and Stock, 2001).

109 Kate Crawford, *Atlas of AI: Power, Politics, and the Planetary Costs of Artificial Intelligence* (New Haven: Yale University Press, 2021).

110 Greenpeace, *Clicking Clean: Who Is Winning the Race to Build a Green Internet?* (Washington, DC: Greenpeace, 2017).

111 Walter Brueggemann, *Sabbath as Resistance: Saying No to the Culture of Now* (Louisville: Westminster John Knox, 2014).

112 Safiya Umoja Noble, *Algorithms of Oppression: How Search Engines Reinforce Racism* (New York: NYU Press, 2018).

113   Paul Scharre, *Army of None: Autonomous Weapons and the Future of War* (New York: W. W. Norton, 2018).

114   Aviv Ovadya and Sam Woolley, "The Coming Dark Age of Information," *Foreign Policy*, January 2019.

115   John C. Havens, *Heartificial Intelligence: Embracing Our Humanity to Maximize Machines* (New York: TarcherPerigee, 2016).

Chapter 9

116   Letty M. Russell, *Church in the Round: Feminist Interpretation of the Church* (Louisville: Westminster John Knox, 1993).

117   Nick Bostrom, *Superintelligence: Paths, Dangers, Strategies* (Oxford: Oxford University Press, 2014).

118   Meredith Broussard, *Artificial Unintelligence: How Computers Misunderstand the World* (Cambridge, MA: MIT Press, 2018).

119   Erik Brynjolfsson and Andrew McAfee, *The Second Machine Age: Work, Progress, and Prosperity in a Time of Brilliant Technologies* (New York: W. W. Norton, 2014).

120   Klaus Schwab, *The Fourth Industrial Revolution* (New York: Crown Business, 2017).

121   Virginia Eubanks, *Automating Inequality: How High-Tech Tools Profile, Police, and Punish the Poor* (New York: St. Martin's Press, 2018).

122   Paul Scharre, *Army of None: Autonomous Weapons and the Future of War* (New York: W. W. Norton, 2018).

123   Peter W. Singer, *Wired for War: The Robotics Revolution and Conflict in the 21st Century* (New York: Penguin Press, 2009).

124   United Nations, *Report of the 2019 Group of Governmental Experts on Lethal Autonomous Weapons Systems* (Geneva: United Nations Office at Geneva, 2019).

125   Nigel Biggar, *In Defence of War* (Oxford: Oxford University Press, 2013).

126   Kate Crawford, *Atlas of AI: Power, Politics, and the Planetary Costs of Artificial Intelligence* (New Haven: Yale University Press, 2021).

127   Greenpeace, *Clicking Clean: Who Is Winning the Race to Build a Green Internet?* (Washington, DC: Greenpeace, 2017).

128   Steven Bouma-Prediger, *For the Beauty of the Earth: A Christian Vision for Creation Care* (Grand Rapids: Baker Academic, 2010).

129   Stuart Russell, *Human Compatible: Artificial Intelligence and the Problem of Control* (New York: Viking, 2019).

130   Allan Dafoe, "AI Governance: A Research Agenda," *Governance of AI Program, Future of Humanity Institute*, University of Oxford, 2018.

131   Brent Daniel Mittelstadt, Patrick Allo, Mariarosaria Taddeo, Sandra Wachter, and Luciano Floridi, "The Ethics of Algorithms: Mapping the Debate," *Big Data & Society* 3, no. 2 (2016): 1–21.

Chapter 10

132   I made this one up! But I made you look :P The aphorism adapts a line attributed to Karl Barth ("Take your Bible and your newspaper, and read both. But interpret newspapers from your Bible."). For discussion, see Karl Barth, *The Word of God and the Word of Man* (New York: Harper, 1957).

133   Jacques Ellul, *The Technological Society* (New York: Vintage, 1964).

134   **Scripture:** Mic. 6:8; Gen. 1:27; Gen. 2:15 (NRSV). (Per Chicago, biblical books are cited in notes, not in the bibliography.)

135   European Commission, *Proposal for a Regulation Laying Down Harmonised Rules on Artificial Intelligence (Artificial Intelligence Act)* (Brussels: European Commission, 2021); The White House, *Blueprint for an AI Bill of Rights* (Washington, DC: OSTP, 2022).

136   Miroslav Volf, *Work in the Spirit: Toward a Theology of Work* (Eugene, OR: Wipf and Stock, 2001).

137   John Dyer, *People of the Screen: How Evangelicals Created the Digital Bible and How It Shapes Their Reading of Scripture* (New York: Oxford University Press, 2023).

138   Brent Daniel Mittelstadt et al., "The Ethics of Algorithms: Mapping the Debate," *Big Data & Society* 3, no. 2 (2016): 1–21.

139   Noreen L. Herzfeld, *In Our Image: Artificial Intelligence and the Human Spirit* (Minneapolis: Fortress, 2002).

140   Tim Hutchings, *Creating Church Online: Ritual, Community and New Media* (New York: Routledge, 2017).

141   Heidi A. Campbell and Stephen Garner, *Networked Theology: Negotiating Faith in Digital Culture* (Grand Rapids: Baker Academic, 2016).

142   Luciano Floridi et al., "AI4People—An Ethical Framework for a Good AI Society," *Minds and Machines* 28, no. 4 (2018): 689–707.

143   UNESCO, *Recommendation on the Ethics of Artificial Intelligence* (Paris: UNESCO, 2021).

144   Kevin P. Kelly, *What Technology Wants* (New York: Viking, 2010), esp. chap. 11 on "technium" and human purpose.

145   Quentin J. Schultze, *Habits of the High-Tech Heart: Living Virtuously in the Information Age* (Grand Rapids: Baker Academic, 2002).

Chapter 2225

146   Lewis Carroll, *Alice's Adventures in Wonderland* (London: Macmillan, 1865).

147   Peter L. Bernstein, *Wedding of the Waters: The Erie Canal and the Making of a Great Nation* (New York: W. W. Norton, 2005).

REFERENCES

148   James Hamilton, *Faraday: The Life* (London: HarperCollins, 2002).

149   Doron Swade, *The Difference Engine: Charles Babbage and the Quest to Build the First Computer* (New York: Viking, 2001).

150   John Emsley, *The 13th Element: The Sordid Tale of Murder, Fire, and Phosphorus* (New York: Wiley, 2000), on friction matches in the 1820s; Leslie Tomory, *Progressive Enlightenment: The Origins of the Gaslight Industry, 1780–1820* (Cambridge, MA: MIT Press, 2012).

151   Roy Porter, *The Greatest Benefit to Mankind: A Medical History of Humanity* (New York: W. W. Norton, 1997); Angus Deaton, *The Great Escape: Health, Wealth, and the Origins of Inequality* (Princeton: Princeton University Press, 2013).

152   Vaclav Smil, *Feeding the World: A Challenge for the Twenty-First Century* (Cambridge, MA: MIT Press, 2000).

153   Vaclav Smil, *Energy and Civilization: A History* (Cambridge, MA: MIT Press, 2017).

154   Arthur Turrell, *The Star Builders: Nuclear Fusion and the Race to Power the Planet* (New York: Scribner, 2021).

155   Nita A. Farahany, *The Battle for Your Brain: Defending the Right to Think Freely in the Age of Neurotechnology* (New York: St. Martin's Press, 2023).

156   Robert Zubrin, *The Case for Mars: The Plan to Settle the Red Planet and Why We Must* (New York: Free Press, 1996); Gerard K. O'Neill, *The High Frontier: Human Colonies in Space* (New York: William Morrow, 1976).

157   Joe Quirk and Patri Friedman, *Seasteading: How Floating Nations Will Restore the Environment, Enrich the Poor, Cure the Sick, and Liberate Humanity from Politicians* (New York: Free Press, 2017).

158   George M. Church and Ed Regis, *Regenesis: How Synthetic Biology Will Reinvent Nature and Ourselves* (New York: Basic Books, 2012); K. Eric Drexler, *Engines of Creation: The Coming Era of Nanotechnology* (New York: Anchor, 1986).159 Jeremy Rifkin, *The Zero Marginal Cost Society: The Internet of Things, the*

Collaborative Commons, and the Eclipse of Capitalism (New York: St. Martin's Press, 2014); Paul Mason, *PostCapitalism: A Guide to Our Future* (New York: Farrar, Straus and Giroux, 2016).

160   Caleb Scharf, *The Copernicus Complex: Our Cosmic Significance in a Universe of Planets and Probabilities* (New York: Scientific American/Farrar, Straus and Giroux, 2014).

161   Stuart Russell, *Human Compatible: Artificial Intelligence and the Problem of Control* (New York: Viking, 2019).

About the Author

The Rev. Dr. Christopher J. Benek is an internationally recognized pastor, theologian, and thought leader at the crossroads of faith and technology. His insights on artificial intelligence and theology have been featured in major outlets including The New York Times, The Guardian, BBC, and NPR, and he has advised companies, policymakers, and governments on AI's ethical and societal impact.

He holds advanced degrees from Princeton and Pittsburgh Theological Seminaries and is pursuing a Ph.D. at Durham University, where his research explores technological futurism and Christian eschatology. Dr. Benek is also the co-founder and founding Board Chair of the Christian Transhumanist Association.

With more than 20 years of pastoral ministry, he currently serves as pastor of First Miami Presbyterian Church and CEO of The CoCreators Network, an organization dedicated to the vision of Better People, Better Tech, Better World.